T0326648

Differential Topology and Geometry with Applications to Physics

Differential Topology and Geometry with Applications to Physics

Eduardo Nahmad-Achar

National Autonomous University of Mexico, Mexico City, Mexico

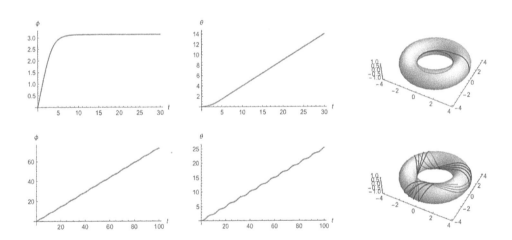

IOP Publishing, Bristol, UK

Permission to make use of IOP Publishing content other than as set out above may be sought at permissions@iop.org.

Eduardo Nahmad-Achar has asserted his right to be identified as the author of this work in accordance with sections 77 and 78 of the Copyright, Designs and Patents Act 1988.

ISBN   978-0-7503-2072-6 (ebook)
ISBN   978-0-7503-2070-2 (print)
ISBN   978-0-7503-2071-9 (mobi)

DOI   10.1088/2053-2563/aadf65

Version: 20181201

IOP Expanding Physics
ISSN 2053-2563 (online)
ISSN 2054-7315 (print)

British Library Cataloguing-in-Publication Data: A catalogue record for this book is available from the British Library.

Published by IOP Publishing, wholly owned by The Institute of Physics, London

IOP Publishing, Temple Circus, Temple Way, Bristol, BS1 6HG, UK

US Office: IOP Publishing, Inc., 190 North Independence Mall West, Suite 601, Philadelphia, PA 19106, USA

To Alex, Luis, and Adela.

Contents

16   Newtonian space–time and thermodynamics   16-1

17   Special relativity, electrodynamics, and the Poincaré group   17-1

18   General relativity   18-1

Preface

Differential geometry has encountered numerous applications in physics. More and more, physical concepts considered as *fundamental* are understood as direct consequences of geometric principles. The mathematical structures of Maxwell's electrodynamics, the general theory of relativity, string theory, and gauge theories, to name but a few, are geometric in nature. All of these disciplines require a curved space for the description of a system, and we require a mathematical formalism that can handle the dynamics in such spaces if we wish to go beyond a simple and superficial discussion of physical relationships. This formalism is precisely *differential geometry*. Even areas such as thermodynamics and fluid mechanics greatly benefit from a differential geometric treatment.

Differential geometry has effected important changes not only in physics, but in important branches of mathematics. Its applications have been so varied and extensive that, on the one hand, many and large (and sometimes complicated) treatments have been written about them and, on the other, many treatments are dispersed in the specialist literature. The consequence of this is that most scientists are not familiar with differential geometry despite the great desirability that they should be.

This book has the purpose of presenting, in a concise and direct manner while maintaining the appropriate mathematical formalism, the fundamentals of differential topology and differential geometry together with essential applications in many branches of physics. In particular, even though the only requirements for reading this book are linear algebra and differential and integral calculus, we arrive at the proof of Stokes' theorem on manifolds and an understanding of the fundamental expressions of advanced calculus in terms of differential forms, and take a good glance at the boundary between differential geometry and algebraic topology, and, on the physics side, at formulating Newtonian mechanics, Maxwell's electrodynamics, and Einstein's general theory of relativity in a geometric language (apart from some applications to mechanics, fluid dynamics, and thermodynamics). In terms of physics it is assumed that the reader is familiar with the elements of the special theory of relativity. The material is presented at an advanced undergraduate and starting graduate level, and in such a way that a specific topic can be learnt in a minimum of time without sacrificing understanding, important concepts, or mathematical formalism.

The book is organised as follows: after describing the need for working on curved spaces (chapter 1) we introduce (chapter 2) the theory of curves and surfaces in $\mathbb{E}^3$ to familiarise the reader with some fundamental concepts such as *curvature* and *torsion*, which will be studied further ahead in an abstract manner and for manifolds of arbitrary dimension. In chapter 3 the elements of set topology are given, and in chapters 4–7 we introduce differentiable manifolds and build tangent spaces, and vector and tensor fields on them. Chapter 8 introduces the reader to exterior calculus and orientable manifolds in preparation for (having first defined maps between manifolds in chapter 9) integration on manifolds and Stokes' theorem (chapter 10).

Chapters 11–13 study Lie derivatives, linear connections, and geodesics, and chapter 14 covers torsion and curvature on manifolds.

We have decided to postpone introducing a metric as much as possible (until chapter 15). This is not usual in physics-oriented books, but it allows the reader to see all that is possible to construct and study on a manifold without the aid of a metric, and eliminates the false idea that many of the structures introduced earlier depend on the existence of a metric.

Chapters 16–18 make use of all the previously introduced formalisms to present, in that language, Newtonian physics, electrodynamics, special relativity, and general relativity. In particular, Einstein's equations are derived from a variational principle and the significance and properties of the energy–momentum tensor are discussed. Finally, chapter 19 deals with first-order perturbations to the metric and gravitational radiation.

An Analytical Index for this eBook may be downloaded from: https://sigi. nucleares.unam.mx/sgiicn/people/user/view/id/43?

Notation

$\mathbb{N}$	The set of natural numbers $\{1, 2, \ldots\}$.
$\mathbb{Z}$	The set of integers.
$\mathbb{R}$	The set of real numbers.
$\mathbb{I}$	The set of imaginary numbers.
$M, N, \ldots$	Topological space; differentiable manifold.
$V^c = M \backslash V$	Complement of V in M, for $V \subset M$.
$\mathcal{P}(M) = 2^M$	Power set of M, i.e. the set of all possible subsets of M.
τ	Topology of a topological space; torsion function of a curve.
κ	Curvature function of a curve.
Φ, Ψ	Atlas of a manifold.
ϕ, ψ	Chart of an atlas.
X_p, Y_p, Z_p	Tangent vector to a manifold at point p.
X, Y, Z	Vector field on a manifold.
α, β, γ	Curve on a manifold.
$\alpha'(t), \dot{\alpha}(t)$	$d\alpha/dt$.
λ, μ, ν	Differential form.
$T_p(M)$	Tangent space to manifold M at point p.
$T_p^*(M)$	Cotangent space to manifold M at point p.
$T(M)$	Tangent bundle to manifold M.
$T^*(M)$	Cotangent bundle to manifold M.
$\mathfrak{X}(M)$	Space of vector fields on manifold M.
$\mathfrak{X}^*(M)$	Space of 1-form fields on manifold M.
h_*	Differential map.
h^*	Pullback.
$\Lambda^k(M)$	Vector space of all k-forms on manifold M.
$\mathcal{G}(M)$	Grassmann algebra of manifold M.
$\Gamma^i_{\ jk}$	Connection coefficients; Christoffel symbols.
$\gamma^i_{\ jk}$	Structure functions.
g_{ab}	Metric tensor.
η_{ab}	Minkowski metric tensor.
$T^k_{\ ij}$	Torsion tensor.
$R^a_{\ bcd}$	Riemann tensor.
R_{ab}	Ricci tensor.
R	Ricci scalar.
G_{ab}	Einstein tensor.
T_{ab}	Energy–momentum tensor.
t^{ik}_{LL}	Landau–Lifshitz energy–momentum pseudo-tensor for the gravitational field.
F_{ab}	Maxwell's electromagnetic tensor.
$D_{\alpha\beta}$	Mass quadrupolar moment.
S_g	Action for the gravitational field.
S_m	Action for the matter fields.
$\epsilon_{\alpha\beta\gamma}$	Levi-Civita symbol.
d	Exterior derivative operator.
$\pounds_X$	Lie derivative with respect to the vector field X.
∇	Covariant derivative; gradient operator.
$\langle \, , \, \rangle_p$	Inner product in $T_p(M)$.

S	Shape operator of a surface.
U	Orientation on $M \subset \mathbb{E}^3$; normal vector field to $M \subset \mathbb{E}^3$.
$\mathrm{supp}(f)$	Support of a function f, i.e. the subset of f's domain where $f \neq 0$.
$\mathrm{img}\, d^{(k)}$	Image of $d^{(k)} : \Lambda^k(M) \rightarrow \Lambda^{k+1}(M)$, i.e. the set of all exact $(k+1)$-forms.
$\ker d^{(k)}$	Kernel of $d^{(k)} : \Lambda^k(M) \rightarrow \Lambda^{k+1}(M)$, i.e. the set of all closed k-forms.
$H^k(M)$	kth de Rham cohomology group for manifold M.
$H_k(M)$	kth homology group for manifold M.
β^k	Betti number.
$\equiv$	Exactly equal to.
$\overset{\mathrm{def}}{=}$	Equal by definition.
$\overset{(6.12)}{=}$	Equal from equation (6.12).
$X \perp Y$	Vector field X is orthogonal to vector field Y.

Acknowledgements

First and foremost I would like to thank my son, Alexander Nahmad-Rohen, who undertook the task of reading the whole manuscript, checking most mathematical developments, translating my notes into English, typesetting them in LaTeX, and making an enormous number of suggestions for improvements along the way. His enthusiasm for understanding the whole subject and his encouragement to make me write this book knew no bounds.

I also wish to thank Henrik Fabian Jendle, who made the first typesetting in LaTeX of the Spanish version of my notes, and several of the figures from my poorly drawn sketches; and Aline Guevara Villegas, for turning the figures in chapters 1–14 into coloured pieces that enhance the presentation of the material. The rest of the figures are my own. Renato Lemus helped improve figure 18.1. Many of the graphs were rendered using *Mathematica*® (Wolfram Research Inc.).

In writing this manuscript I have benefited from teaching the course 'Differential Topology and Geometry with Applications to Physics' at the National University of Mexico (UNAM) to advanced undergraduate and beginning graduate students, most of whom have expressed their desire to see my notes written as a book. I am very grateful to the publishers at IOP for making this possible with excellent craftsmanship.

Author biography

Eduardo Nahmad-Achar

 Eduardo Nahmad-Achar earned his BSc in Physics and BSc in Mathematics from the National University of Mexico, and later his MSc in Applied Mathematics and PhD in Physics from the University of Cambridge, UK. He is the author of many scientific publications and has been invited to international conferences to talk about his achievements. He has lectured extensively at UNAM in various topics of physics and mathematics, including differential geometry, general relativity, advanced mathematics, quantum information, and quantum physics, at both graduate and undergraduate levels. He was also Founding Director of the Centre for Polymer Research, nr. Mexico City.

IOP Publishing

Differential Topology and Geometry with Applications to Physics

Eduardo Nahmad-Achar

Chapter 1

Synopsis of general relativity

This chapter will describe the need to study the geometry of curved spaces, how to make measurements on them, and the dynamics of those measurements. In subsequent chapters we shall build the analytic tools necessary to study the geometry of curved space and also show several applications.

The *general theory of relativity* is the extension of the *special theory of relativity* to accelerated reference frames. Galileo's experiments showed that the acceleration on an object caused by a gravitational field is independent of the object. To any observer, then, an accelerated reference frame is equivalent to an inertial frame in the presence of a gravitational field. In other words, an arbitrary observer cannot distinguish between standing on Earth and being in a windowless rocket travelling upwards and accelerating at a rate equivalent to Earth's gravitational pull ($\approx$9.8 m s^{-2}), and floating in space far from any object is equivalent to free fall under the effect of Earth's gravitational field, as illustrated in figure 1.1.

General relativity becomes therefore a theory of gravity, and in this context the idea of an 'inertial reference frame' loses its meaning.

If there are no longer any privileged reference frames, the physics studied in any one frame must be equivalent to that studied in any other and the equations of physics (if written *appropriately*, i.e. independent of coordinates) must be the same in all frames. It is thus necessary to extend the group of Lorentz transformations (which describe, according to special relativity, the way in which two inertial observers' space–time coordinates are related to each other) to the group of *all* continuous coordinate transformations (linear and nonlinear). In other words, *the equations which describe natural phenomena must be covariant with respect to all continuous coordinate transformations*. This statement is known as the 'principle of relativity'.

Under this group of transformations, the only thing that remains invariant is the fact that two points which are very close to each other have very similar coordinates (see figure 1.2).

doi:10.1088/2053-2563/aadf65ch1

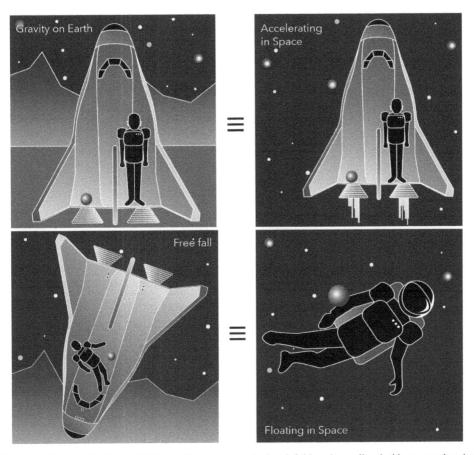

Figure 1.1. Top: equivalence of being subject to a gravitational field and standing inside an accelerating rocket, with the same acceleration as the field. Bottom: equivalence of being in free fall and floating in space.

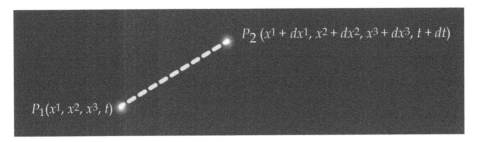

Figure 1.2. Points that are close to each other have similar coordinates, regardless of any continuous transformation in space.

Suppose we send a light ray from P_1 to P_2. Special relativity tells us that the speed of light, c, is constant regardless of the observer. Thus,

$$\sqrt{dx^2 + dy^2 + dz^2} = c\, dt$$

i.e. $c^2 dt^2 - dx^2 - dy^2 - dz^2 = 0$, which must be satisfied for any two points which are close to each other. The quantity

$$
\begin{aligned}
\Delta\tau^2 &= \Delta t^2 - \frac{\Delta x^2 + \Delta y^2 + \Delta z^2}{c^2} \\
&= \Delta t^2 - \frac{|\Delta r^2|}{c^2}
\end{aligned}
\tag{1.1}
$$

is therefore invariant under coordinate transformations because c is the same constant in all reference frames. This result is obtained in special relativity and is therefore locally valid in any reference frame.

τ is analogous to the arc length parameter of a curve on $\mathbb{R}^3$. In other words, if a curve is given by $x^i = x^i(s)$, $i \in \{1,2,3\}$ then the arc length Δs between two nearby points x^i and $x^i + \Delta x^i$ of the curve is given by:

$$\Delta s^2 = (\Delta x^1)^2 + (\Delta x^2)^2 + (\Delta x^3)^2 \text{ in Cartesian coordinates,}$$

$$\Delta s^2 = \Delta r^2 + r^2(\Delta\theta^2 + \sin\theta^2 \Delta\phi^2) \text{ in spherical coordinates,}$$

as we shall see in chapter 2. We shall also see that a curve parametrised by its arc length has unitary speed, i.e. s can be thought of as its 'proper time'. This suggests the following definition.

1.1 Definition (proper time)

τ is called the 'proper time' of the path from P_1 to P_2.

If a particle travels from P_1 to P_2, τ is the natural time parameter a clock travelling with the particle would measure. Proper time, defined in such a way, must be invariant because it has been defined without reference to any coordinate system. We have seen that if the particle travels along a light ray's path then $\Delta\tau = 0$; along other paths, $\Delta\tau$ will take on different values.

1.2 Remark

Equation (1.1) is rather powerful; in fact, we may consider it to be the basis of special relativity, deriving from it the Lorentz transformations and other consequences (see examples 1.3 and 1.5).

1.3 Example (time dilation)

Consider two events on a body's path (see figure 1.3). The body moves with velocity v with respect to a coordinate system. Then $|\Delta r| = v\Delta t$, where Δr is the travelled distance and Δt is the elapsed time.

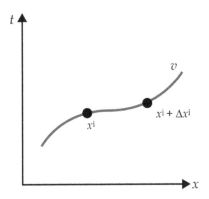

Figure 1.3. Two events, x^i and $x^i + \Delta x^i$, along the path of a body moving with velocity v with respect to a coordinate system.

From equation (1.1),

$$\Delta\tau^2 = \Delta t^2 - \frac{|\Delta r^2|}{c^2} = \Delta t^2 - \Delta t^2\left(\frac{v}{c}\right)^2$$

or

$$\Delta t = \frac{\Delta\tau}{\sqrt{1 - \left(\dfrac{v}{c}\right)^2}}$$

But $\Delta\tau$ is the time measured by the body's clock. Therefore, the coordinate time interval Δt measured in a reference frame in which the body is moving is greater than the proper time interval $\Delta\tau$ measured by a clock at rest with respect to the body. In other words, moving clocks tick more slowly.

1.4 Definition (proper distance)

If $|\Delta r| > c\,\Delta t$, then $\Delta\tau^2 < 0$ and therefore $\Delta\tau$ would be imaginary ($\Delta\tau \in \mathbb{I}$). Thus, the proper time interval between two events for which $|\Delta r| > c\,\Delta t$ in a reference frame is imaginary (i.e. for a particle moving from one event to the other an imaginary amount of time will have elapsed). For this to happen, the particle must move with velocity $v > c$, which is impossible, and therefore the two aforementioned events cannot both be in a particle's path.

Now, $c\sqrt{(-\Delta\tau)^2} > 0$. Moreover, $c\sqrt{(-\Delta\tau)^2}$ has units of distance. This quantity is called the 'proper distance' between the two events. It is the distance which would be measured by a standard ruler in a reference frame where the events occurred simultaneously, for if $\Delta t = 0$ then $|\Delta r^2| = -c^2\Delta t^2$, i.e. $\Delta r = c\sqrt{(-\Delta\tau)^2}$.

1.5 Example (Lorentz contraction)

Suppose a reference frame F' is moving with respect to another reference frame, F, with velocity v in direction x, as shown figure 1.4.

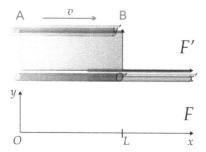

Figure 1.4. A reference frame F' moving with respect to another reference frame F. As seen by an observer on F', the length of an object on F would appear contracted in the direction of the movement.

Consider the following two events, and let us see how they relate:
- Event A: O and O' coincide.
- Event B: L and O' coincide.

In F,

$$\Delta\tau^2 = \Delta t^2 - \frac{|\Delta r^2|}{c^2} = \frac{L^2}{v^2} - \frac{L^2}{c^2}$$

In F', both events occur at the origin, O'. Therefore, $\Delta r' = 0$ and

$$\Delta\tau^2 = \Delta t'^2 - \frac{|\Delta r'^2|}{c^2} = \frac{L'^2}{v^2} - 0 = \frac{L'^2}{v^2}$$

where $\frac{L'}{v}$ is the distance between O and L as measured in F' divided by the velocity with which O' passes O and L. From the above expressions we have

$$L' = L\sqrt{\left(1 - \frac{v^2}{c^2}\right)}$$

In other words, a bar would appear to be contracted ($L' < L$) if its length were measured in a reference frame in which it is not at rest.

1.6 Coordinate transformation

Under a general continuous coordinate transformation, $(t,\,x,\,y,\,z)\longmapsto(x^0,\,x^1,\,x^2,\,x^3)$, as pictured in figure 1.5, we have

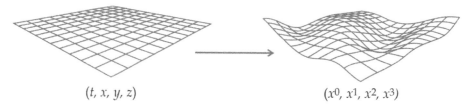

$(t,\,x,\,y,\,z)$ $(x^0,\,x^1,\,x^2,\,x^3)$

Figure 1.5. A general continuous coordinate transformation.

$$t = t(x^0, x^1, x^2, x^3)$$
$$x = x(x^0, x^1, x^2, x^3)$$
$$y = y(x^0, x^1, x^2, x^3)$$
$$z = z(x^0, x^1, x^2, x^3)$$

$$\Delta t = \frac{\partial t}{\partial x^0} \Delta x^0 + \frac{\partial t}{\partial x^1} \Delta x^1 + \frac{\partial t}{\partial x^2} \Delta x^2 + \frac{\partial t}{\partial x^3} \Delta x^3$$
$$\Delta x = \frac{\partial x}{\partial x^0} \Delta x^0 + \frac{\partial x}{\partial x^1} \Delta x^1 + \frac{\partial x}{\partial x^2} \Delta x^2 + \frac{\partial x}{\partial x^3} \Delta x^3$$
$$\vdots$$

Substituting in equation (1.1) and setting $c = 1$,

$$\Delta\tau^2 = \left[\left(\frac{\partial t}{\partial x^0}\right)^2 - \left(\frac{\partial x}{\partial x^0}\right)^2 - \left(\frac{\partial y}{\partial x^0}\right)^2 - \left(\frac{\partial z}{\partial x^0}\right)^2 \right](\Delta x^0)^2 + \cdots$$

$$+ 2\left[\frac{\partial t}{\partial x^0}\frac{\partial t}{\partial x^1} - \frac{\partial x}{\partial x^0}\frac{\partial x}{\partial x^1} - \frac{\partial y}{\partial x^0}\frac{\partial y}{\partial x^1} - \frac{\partial z}{\partial x^0}\frac{\partial z}{\partial x^1} \right]\Delta x^0 \Delta x^1 + \cdots$$

or

$$\Delta\tau^2 = \sum_a \sum_b g_{ab}\Delta x^a \Delta x^b \quad (a, b \in \{0, 1, 2, 3\}) \tag{1.2}$$

where

$$g_{ab} \stackrel{\text{def}}{=} \frac{\partial t}{\partial x^a}\frac{\partial t}{\partial x^b} - \frac{\partial x}{\partial x^a}\frac{\partial x}{\partial x^b} - \frac{\partial y}{\partial x^a}\frac{\partial y}{\partial x^b} - \frac{\partial z}{\partial x^a}\frac{\partial z}{\partial x^b} \tag{1.3}$$

It is common to use the 'Einstein notation', in which repeated index variables imply an implicit sum:

$$g_{ab}\Delta x^a \Delta x^b \stackrel{\text{def}}{=} \sum_a \sum_b g_{ab}\Delta x^a \Delta x^b$$

We may then write

$$\Delta\tau^2 = g_{ab}\Delta x^a \Delta x^b \tag{1.4}$$

Implicit in the quantity g_{ab} is the fact that *points close to each other in a reference frame remain close to each other under a coordinate transformation*. g_{ab} tells us, then, how to make measurements in a given reference frame: more precisely (see equation (1.2)), it tells us how to obtain the proper time interval between two points given the difference in their coordinates. This interval, $\Delta\tau$, is invariant; the difference in coordinates is not.

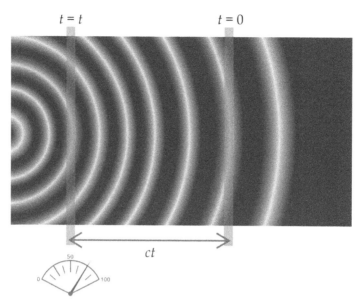

Figure 1.6. A light wave travelling towards the right passes a frequency detector at rest.

1.7 The Doppler effect

Figure 1.6 shows a light wave travelling towards the right.

A detector at rest measures the light's frequency, f, by measuring the number of crests which pass it within a time interval t. Thicker lines indicate the crests which pass the detector at time $t = 0$ and at time $t > 0$. If N crests passed the detector during that time interval, we have $f = \frac{N}{t}$.

However, only N' crests will pass a detector which moves towards the right with velocity v, as shown in figure 1.7.

Since

$$\frac{N'}{N} = \frac{(c-v)t}{c\,t} = 1 - \frac{v}{c}, \quad \text{we have} \quad f' = \frac{N'}{t} = \frac{N}{t}\left(1 - \frac{v}{c}\right) = \left(1 - \frac{v}{c}\right)f,$$

so the detector moving away from the light source measures a smaller frequency. This effect is called the 'Doppler effect' or 'redshift', since red light has a lower frequency than the rest of the visible spectrum.

1.8 Gravitational redshift

Suppose a photon is emitted at time $t = 0$ by a light source, as depicted in figure 1.8. The photon has frequency f_0 and travels upwards from the ground. An observer A standing atop a building lets himself fall towards the ground with initial velocity $v = 0$ at that same instant. A second observer, B, remains on top of the building.

Observer A is in free fall (which is equivalent to floating in space), and therefore he measures the photon's frequency as $f_A = f_0$. With respect to observer A, observer

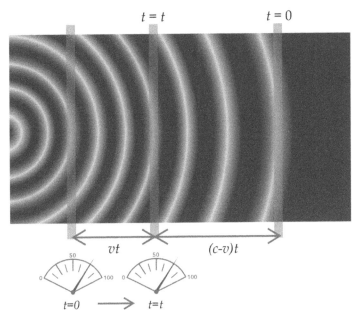

Figure 1.7. A light wave travelling towards the right passes a frequency detector at rest and a frequency detector moving towards the right with velocity v.

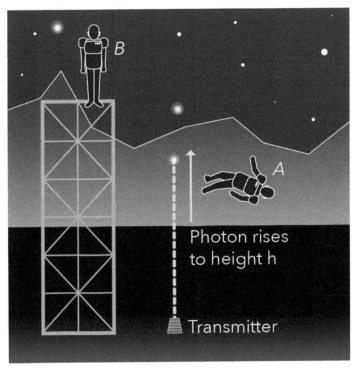

Figure 1.8. A photon of frequency f_0 is emitted upwards from the ground as an observer atop a building lets himself fall towards the ground.

B is moving away from the light source with velocity v at a given time. Therefore, he measures $f_B = (1 - \frac{v}{c})f_A = (1 - \frac{v}{c})f_0$.

Suppose observer A and the photon meet at height h at a time $t = \frac{h}{c}$. At that time, observer A, subject to an acceleration g, has acquired a velocity $gt = g(\frac{h}{c})$. Thus,

$$f_B = f_0\left(1 - \frac{g\,h}{c^2}\right) < f_0$$

1.9 Curved space, Schild's argument

Consider two observers ϑ_1 and ϑ_2 at rest with respect to Earth and to each other. Observer ϑ_1 is standing at a height z_1 above ground level, while observer ϑ_2 is standing at a height $z_2 = z_1 + h$ $(h > 0)$.

Suppose observer ϑ_1 emits an electromagnetic signal towards observer ϑ_2. The signal consists of a single pulse of exactly N cycles with frequency ν_1. The time interval required for the emission is $\Delta t_1 = \frac{N}{\nu_1}$. Observer ϑ_2 receives the pulse and measures its frequency as ν_2 during a time interval $\Delta t_2 = \frac{N}{\nu_2}$. But we have seen that $\nu_2 < \nu_1$, so we must necessarily have $\Delta t_2 > \Delta t_1$.

In a flat space, even considering the effects of Earth's gravitational field on light rays, the paths of both ends of the pulse must be congruent (see figure 1.9), resulting in a parallelogram with two opposite sides of unequal length (!). Therefore, in the presence of a gravitational field, $\Delta t_2 > \Delta t_1$ means that *space must be curved*.

1.10 On the dynamics of g_{ab}

All we have assumed so far is that the acceleration caused by a gravitational field on an object is independent of the object (Galileo's principle). Under this assumption, being in the presence of a gravitational field g is equivalent to being in a curved space. g_{ab} measures how space is curved. Therefore, g_{ab} is a measure of g.

It is therefore important to ask what the dynamics of g_{ab} are.

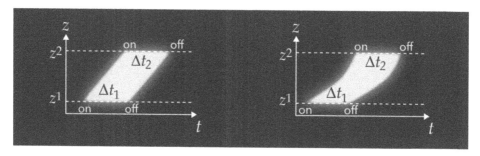

Figure 1.9. A light pulse of duration Δt_1 is emitted by an observer upwards to another observer (at rest with the first one) who receives it during a time interval Δt_2, which must necessarily be larger in the presence of a gravitational field. The paths on both ends of the pulse must be congruent, which leads to a contradiction if space–time is flat.

Any differential equation postulated for g_{ab} must be of second or higher order. Assuming it does not contain derivatives of higher-than-second order with respect to the coordinates, the principle of relativity, Gauss's metric theory of surfaces and its extension to arbitrary dimension by Riemann imply, as we shall later see, that

$$R_{ab} = 0 \quad \text{(in vacuum)}$$

$$R_{ab} - \frac{1}{2}Rg_{ab} = 8\pi T_{ab} \quad \text{(in general)} \tag{1.5}$$

where $R^m{}_{abn}$ is the curvature of space–time, $R_{ab} = R^m{}_{amb}$, $R = R^a{}_a$, and T_{ab} is the stress–energy tensor of the non-gravitational fields.

We shall also see that equation (1.5) may be obtained from a variational principle where the Lagrangian is of the form $L = \sqrt{-g}\,R$. In order to understand this, we must first introduce enough mathematical formalism to be able to perform mathematical analysis on curved spaces. We will do this in the following chapters.

1.11 Geometric units

Through general relativity, we understand gravitation in purely geometric terms. It is then natural to express physical quantities in units of length.

This can be achieved by using a unit system in which both the speed of light, c, and Newton's gravitation constant, G, are equal to unity:

$$c = 1 \qquad G = 1$$

Under this convention, there is no longer any distinction between seconds and metres, and neither is there any distinction between kilograms and metres. For instance, 1 s of length equals the distance a photon travels in 1 s of time: approximately 3×10^8 m.

To give ourselves a clearer idea of these 'geometric units', consider the following examples:

(i) 1 m of time $= \dfrac{1\,\text{m}}{3 \times 10^8\,\text{ms}^{-1}} = 3.3 \times 10^{-9}$ s $= 3.3$ ns (the amount of time in which a light ray travels a distance of 1 m).

(ii) 1 cm of mass $= \dfrac{1\,\text{cm}}{0.74 \times 10^{-28}\text{cmg}^{-1}} \simeq 1.4 \times 10^{28}$ g $\simeq$ Earth's mass $\left(\dfrac{1\,\text{cm}}{\left(G/c^2\right)}\right)$.

Whenever we require the use of units and unless stated otherwise, we shall use geometric units.

Chapter 2

Curves and surfaces in $\mathbb{E}^3$

I. Curves

2.1 Curvature, torsion, and Frenet's formulae

The considerations in chapter 1 suggest that, if we wish to study phenomena in the presence of gravitational fields, we must do so in curved spaces. The theory of general relativity describes natural phenomena quite elegantly in terms of the geometry of 'space–time', a *four-dimensional Lorentzian manifold* which represents the continuum in which all events occur spatially and temporally. Before introducing the abstract concept of a differentiable manifold and learning analytic techniques on such a space, it is best to see how to study the geometry of curves and surfaces which are immersed in a flat three-dimensional space, $\mathbb{R}^3$. This will paint a clear mental picture of the meaning of concepts such as curvature and torsion, which will help us when we define said concepts and study them on spaces of higher dimension and on spaces which do not 'live' in $\mathbb{R}^n$.

The area of mathematics which introduces the tools of differential and integral calculus on these more general spaces and studies their geometric properties is called 'differential geometry'. This discipline has allowed for the formulation of many ideas in physics in a simpler way, leading us to a more fundamental understanding of these ideas. We shall demonstrate this in the following chapters.

In this chapter we will study geometric concepts of curves and surfaces. We use the symbol $\mathbb{R}^3$ to denote the Cartesian space $\mathbb{R} \times \mathbb{R} \times \mathbb{R}$ of triads (x, y, z) which we identify with points in space, and the symbol $\mathbb{E}^3$ to denote this same space seen as a vector space and where the (vector and scalar) products between vectors are defined. $\mathbb{E}^3$ is called the 'Euclidean three-dimensional space', and the curves and surfaces we will study 'live' in this space.

As the main purpose here is to create a proper mental picture by using familiar curves and surfaces, we shall follow, in a concise and free manner, the developments of what is called *classical differential geometry* presented in many textbooks, notably Spivak's *A Comprehensive Introduction to Differential Geometry*, vol 2, and O'Neill's *Elementary Differential Geometry*.

2.2 Definition (curve in $\mathbb{E}^3$)

A 'curve in $\mathbb{E}^3$' is a differentiable function

$$\alpha: I \subset \mathbb{R} \longrightarrow \mathbb{E}^3$$
$$t \longmapsto (\alpha_1, \alpha_2, \alpha_3)$$

2.3 Example

   (i) Helix: The curve $t \longmapsto (a \cos t, a \sin t, 0)$ travels along a circle of radius $a > 0$ on the xy plane of $\mathbb{E}^3$. Therefore $\alpha(t) = (a \cos t, a \sin t, b\,t)$ represents a circular helix if $b \neq 0$ (see figure 2.1).

   (ii) Hyperbolic helix: $\alpha(t) = (e^t, e^{-t}, \sqrt{2}\,t)$ also represents a helix, but note that, since $\alpha_1(t) \cdot \alpha_2(t) = 1$, the helix follows a hyperbola on the xy plane (see figure 2.2).

2.4 Definition (velocity of a curve)

Let $\alpha: I \longrightarrow \mathbb{E}^3$ be a curve in $\mathbb{E}^3$, and let $\alpha = (\alpha_1, \alpha_2, \alpha_3)$. For each $t \in I$, the 'velocity vector of α at t' is

$$\alpha'(t) \stackrel{\text{def}}{=} \left(\frac{d}{dt}\alpha_1(t), \frac{d}{dt}\alpha_2(t), \frac{d}{dt}\alpha_3(t) \right)_{\alpha(t)}$$

α is called 'regular' if $\alpha'(t) \neq 0 \; \forall\, t \in I$. The curve and its velocity vector are depicted in figure 2.3.

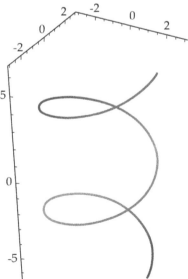

Figure 2.1. A circular helix in $\mathbb{E}^3$.

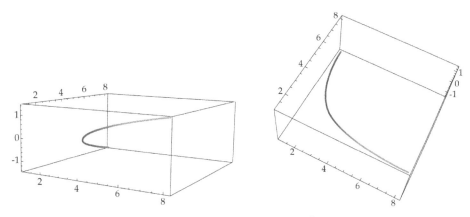

Figure 2.2. A hyperbolic helix in $\mathbb{E}^3$.

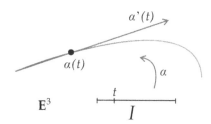

Figure 2.3. A curve in $\mathbb{E}^3$ with its velocity vector.

2.5 Definition (reparametrisation of a curve)

Let $\alpha: I \longrightarrow \mathbb{E}^3$ be a curve in $\mathbb{E}^3$, and let $h: J \subset \mathbb{R} \longrightarrow I$ be a differentiable map. Then

$$\beta = \alpha \circ h: J \longrightarrow \mathbb{E}^3$$
$$q \longmapsto \alpha(h(q))$$

is called a 'reparametrisation' of α by h.

2.6 Lemma

If β is the reparametrisation of α by h, then

$$\beta'(s) = \frac{d}{ds}h(s) \cdot \alpha'(h(s))$$

Proof. Trivial, using the chain rule for derivatives. $\square$

Note that β travels along the same path as α but, in general, reaches a given point p on the path at a time t different from the time at which α reaches p (see figure 2.4).

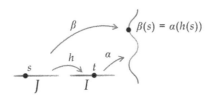

Figure 2.4. The reparametrisation β of curve α by a function h.

2.7 Definition (arc length)

Let $\alpha\colon I \longrightarrow \mathbb{E}^3$ be a curve in $\mathbb{E}^3$, and let $\alpha'(t) = \left(\frac{d\alpha_1}{dt}, \frac{d\alpha_2}{dt}, \frac{d\alpha_3}{dt}\right)$ be the velocity of α. Then

$$v \stackrel{\mathrm{def}}{=} \|\alpha'\| = \left[\left(\frac{d\alpha_1}{dt}\right)^2 + \left(\frac{d\alpha_2}{dt}\right)^2 + \left(\frac{d\alpha_3}{dt}\right)^2\right]^{\frac{1}{2}}$$

is called the 'speed of α at t'. The distance travelled by a moving point is determined by integrating its velocity with respect to time. We then define the 'arc length' of α from $t = a$ to $t = b$ as

$$\int_a^b \|\alpha'(t)\|\, dt$$

Sometimes, especially in geometry, what interests us is the trajectory a curve follows and not the particular speed at which it does so. One way to avoid the influence of a curve's speed is to reparametrise it to a curve β of unit speed: $\|\beta'\| = 1$.

2.8 Theorem (reparametrisation by arc length)

If $\alpha\colon I \longrightarrow \mathbb{E}^3$ is a regular curve in $\mathbb{E}^3$, there exists a parametrisation β of α such that β has unit speed.

Proof. Let $a \in I$, and we define

$$s(t) := \int_a^t \|\alpha'(u)\|\, du$$

Since α is regular,

$$\frac{ds}{dt} = \|\alpha'(t)\| > 0$$

Therefore, s has an inverse function $t = t(s)$ and $\frac{dt}{ds} > 0$.

We define $\beta(s) := \alpha(t(s))$. Then, by lemma 2.6,

$$\beta'(s) = \frac{dt}{ds}\alpha'(t(s))$$

and therefore

$$\|\beta'(s)\| = \frac{dt}{ds}(s)\|\alpha'(t(s))\| = \frac{dt}{ds}(s)\frac{ds}{dt}(t(s)) = 1$$

□

The curve β of unit speed has a 'parametrisation by arc length', since the arc length of β from $s = a$ to $s = b$ $(a \leqslant b)$ is $b - a$.

2.9 Example

Consider the helix (i) from example 2.3: $\alpha(t) = (a \cos t, a \sin t, b\, t)$. Then

$$\alpha'(t) = (-a \sin t, a \cos t, b)$$
$$\|a'(t)\|^2 = \alpha'(t) \cdot \alpha'(t) = a^2 \sin^2 t + a^2 \cos^2 t + b^2 = a^2 + b^2$$

In other words, α has constant speed $c = \|\alpha'\| = (a^2 + b^2)^{\frac{1}{2}}$. Measuring the arc length from $t = 0$, we have

$$s(t) = \int_0^t c\, dt = c\, t$$

Therefore,

$$t(s) = \frac{s}{c}$$

and finally

$$\beta(s) = \alpha\left(\frac{s}{c}\right) = \left(a \cos \frac{s}{c}, a \sin \frac{s}{c}, \frac{b\, s}{c}\right), \qquad \|\beta'(s)\| = 1$$

2.10 Definition (vector field)

A 'vector field' Y on a curve α: $I \longrightarrow \mathbb{E}^3$ is a function which maps each $t \in I$ to a vector $Y(t)$ in the tangent space of $\mathbb{E}^3$ at $\alpha(t)$.

(The concept of *tangent space* shall be introduced later on. In the case of $\mathbb{E}^3$, its tangent space at any point is itself; $Y(t)$ is then a vector at $\alpha(t)$ itself.)

2.11 Example

$\alpha'(t)$ is tangent to α at t. In general, however, it need not be so: the 'acceleration of α', for instance, is given by

$$\alpha''(t) = \left(\frac{d^2\alpha_1}{dt^2}(t), \frac{d^2\alpha_2}{dt^2}(t), \frac{d^2\alpha_3}{dt^2}(t)\right)$$

and is not tangent to α.

2.12 Tangent, normal, binormal and Frenet's system

Let $\beta: I \longrightarrow \mathbb{E}^3$ be a curve parametrised by arc length (i.e. $\|\beta'(s)\| = 1 \; \forall s \in I$). $T = \beta'$ is called the 'unit tangent vector field' on β. Since T has a constant magnitude equal to 1, $T' = \beta''$ measures the way in which β changes direction in $\mathbb{E}^3$. T' is called the 'curvature vector field of β'.

Since $T \cdot T = 1$, differentiating we obtain $2\, T \cdot T' = 0$, which means that $T' \perp T$. In other words, T' is normal to β.

$\kappa(s) \overset{\text{def}}{=} \|T'(s)\|$ is called the 'curvature function of β'. Note that $\kappa \geqslant 0$ and that the larger the value of κ the sharper the curvature of β (i.e. the more pronounced the turns of β in $\mathbb{E}^3$). When $\kappa \neq 0$, we define

$$N := \frac{T'}{\kappa}$$

which gives us the direction in which β turns at each point. N is called the 'principal normal vector field' of β:

$$\|N\| = \frac{1}{\kappa}\, \|T'\| = \frac{1}{\kappa}\, \kappa = 1, \qquad N \perp T$$

$B \overset{\text{def}}{=} T \times N$ is called the 'binormal vector field' of β.

Then T, N, B are unit-length, mutually orthogonal vector fields at each point of β.

$$T = \beta', \qquad N = \frac{1}{\kappa}\, T', \qquad B = T \times N$$

constitute 'Frenet's reference-frame field', which is the natural reference frame for the curve β at each point. Figure 2.5 shows the construction of the Frenet frame for a curve in $\mathbb{E}^3$.

2.13 Theorem (Frenet's formulae)

Let $\beta: I \longrightarrow \mathbb{E}^3$ be a curve parametrised by arc length, with curvature function $\kappa > 0$. Then there is a function τ such that

$$T' = +\kappa\, N$$
$$N' = -\kappa\, T + \tau\, B$$
$$B' = -\tau\, N$$

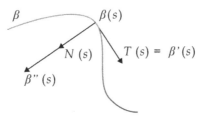

Figure 2.5. Construction of the Frenet frame for a curve in E^3.

Proof. The first equation follows from the definition of κ.

Now, $B \cdot B = 1$ and therefore $B' \cdot B = 0$, which means that $B' \perp B$. Since, by definition, $T' = \kappa N$ and $B = T \times N$, we have $B \cdot T = 0$, which means that $B' \cdot T + B \cdot T' = 0$ or, in other words, $B' \cdot T = -B \cdot T' = -B \cdot \kappa N = 0$. Therefore, $B' \perp T$ and B is a multiple of N: there exists $\tau = \tau(s)$ such that $B' = -\tau N \ \forall s \in I$, rendering the third equation true.

To prove the second equation, we expand N' in the basis (T, N, B):

$$N' = (N' \cdot T)T + (N' \cdot N)N + (N' \cdot B)B$$

(i) since $N \cdot T = 0$, $N' \cdot T + N \cdot T' = 0$, and therefore $N' \cdot T = -N \cdot T' = -N \cdot \kappa N = -\kappa$;

(ii) since $N \cdot N = 1$, $N' \cdot N = 0$;

(iii) since $N \cdot B = 0$, $N' \cdot B = -N \cdot B' = -N \cdot (-\tau N) = \tau$.

$\square$

2.14 Definition (curvature and torsion)

κ is called the 'curvature function' of β. τ is called the 'torsion function' of β.

2.15 Example

Consider the unit-speed helix (see example 2.9):

$$\beta(s) = \left(a \cos \frac{s}{c}, \ a \sin \frac{s}{c}, \ \frac{b \, s}{c}\right)$$

where $a > 0$ and $c = \sqrt{a^2 + b^2}$.

$$T(s) = \beta'(s) = \left(-\frac{a}{c} \sin \frac{s}{c}, \ \frac{a}{c} \cos \frac{s}{c}, \ \frac{b}{c}\right)$$

therefore

$$T'(s) = \left(-\frac{a}{c^2} \cos \frac{s}{c}, \ -\frac{a}{c^2} \sin \frac{s}{c}, \ 0\right), \qquad \kappa(s) = \|T'(s)\| = \frac{a}{c^2} = \frac{a}{a^2 + b^2} > 0$$

and, since $T' = \kappa N$,

$$N(s) = \left(-\cos \frac{s}{c}, \ -\sin \frac{s}{c}, \ 0\right)$$

Likewise, $B = T \times N$, so

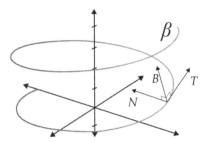

Figure 2.6. Frenet frame for the helix in example 2.9, from which we calculate its curvature and torsion.

$$B(s) = \left(\frac{b}{c} \sin \frac{s}{c}, -\frac{b}{c} \cos \frac{s}{c}, \frac{a}{c} \right), \qquad B'(s) = \left(\frac{b}{c^2} \cos \frac{s}{c}, \frac{b}{c^2} \sin \frac{s}{c}, 0 \right)$$

and $B' = -\tau N$, so

$$\tau(s) = \frac{b}{c^2} = \frac{b}{a^2 + b^2}$$

Observe that
  (i) the helix has constant curvature and torsion;
  (ii) N always points directly towards the helix's axis (see figure 2.6);
  (iii) if $b = 0$, the helix becomes a circle of radius a and the curvature of the circle is $\frac{1}{a}$.

2.16 Lemma (plane curve)

Let $\beta: I \longrightarrow \mathbb{E}^3$ be a curve parametrised by arc length with $\kappa > 0$. Then β is a plane curve (i.e. it lies entirely in a plane) if and only if $\tau = 0$.

Proof. Let us first assume that β is a plane curve. Let us call the plane where it lies Π. Since any plane may be described using a point contained in it and a vector orthogonal to the plane, there exist $p \in \Pi$ and $\vec{q}$ such that $\overrightarrow{r - p} \cdot \vec{q} = 0$ for all points $r \in \Pi$. This is true in particular for all points in β: $(\beta(s) - p) \cdot \vec{q} = 0 \ \forall s \in I$. Taking the first and second derivatives of both sides of this equation, $\beta'(s) \cdot \vec{q} = \beta''(s) \cdot \vec{q} = 0$. Therefore,

$$\vec{q} \perp T = \beta', \qquad \vec{q} \perp N = \frac{\beta''}{\kappa}$$

But B is also orthogonal to T and to N and has unit magnitude, so

$$B = \pm \frac{\vec{q}}{\|\vec{q}\|}$$

In other words, $B' = 0$ and therefore $\tau = 0$.

Now let us assume that $\tau = 0$. Then $B' = 0$, which means that B is a vector field which points in the same direction everywhere. We define $f(s) := (\beta(s) - \beta(0)) \cdot B$

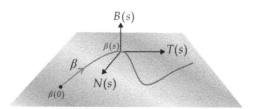

Figure 2.7. A curve with null torsion lies on a plane.

for all $s \in I$. Taking the derivative, $\frac{df}{ds} = \beta' \cdot B = T \cdot B = 0$. But $f(0) = 0$, so $f(s) \equiv 0 \ \forall s \in I$.

   In other words, $(\beta(s) - \beta(0)) \cdot B \equiv 0$ and, therefore, all points of β lie on a plane perpendicular to B, as shown in figure 2.7. $\qquad\square$

2.17 Exercise (plane circle)

Prove that if $\kappa > 0$ is constant and $\tau = 0$ then $\beta(s)$ is part of a circle of radius $\frac{1}{\kappa}$.

Proof. $\tau = 0$ implies that β is a plane curve (see lemma 2.16). We therefore need only to prove that all points of β lie at a distance $1/\kappa$ from a fixed point. We define the curve

$$\gamma := \beta + \frac{1}{\kappa} N$$

Then

$$\gamma' \overset{\kappa=\text{const.}}{=} \beta' + \frac{1}{\kappa} N' = T + \frac{1}{\kappa}(-\kappa \, T) \overset{\tau=0}{=} 0$$

In other words, γ is constant: $\beta(s) + \frac{1}{\kappa} N(s) = \gamma(s) \equiv c$. The distance from c to $\beta(s)$ is

$$\|c - \beta(s)\| = \left\| \frac{1}{\kappa} N(s) \right\| = \frac{1}{\kappa}$$

so $\beta(s)$ is part of a circle of radius $\frac{1}{\kappa}$ centred on point c. $\qquad\square$

II. Surfaces

2.18 Remark

We shall consider, in what follows, regular surfaces $M \subset \mathbb{E}^3$ whose tangent plane $T_p(M)$ is well defined at each point $p \in M$. The tangent plane $T_p(M)$ is a copy of $\mathbb{E}^2$ with its origin at p, and it consists of all vectors v which are tangent to M at point p. Since multiplying a vector of $T_p(M)$ by a scalar and adding two vectors of $T_p(M)$

both result in another vector of $T_p(M)$, the tangent plane at each point is a two-dimensional vector space.

Given $v_p \in T_p(M)$, the function $t \longmapsto p + t\, v_p$ defines a straight line in $\mathbb{E}^3$ tangent to M at p, and if $f: \mathbb{E}^3 \longrightarrow \mathbb{R}$ is an arbitrary function then the application

$$v_p[f] = \frac{d}{dt}f(p + t\, v_p)\Big|_{t=0}$$

tells us how f changes as we move along this line, at the point p. In other words, it is the 'directional derivative' of f along v_p at p.

We then have two ways of looking at a vector tangent to a surface: as an object with magnitude and direction defined at a point $p \in M$, and as a linear operator over functions with real values

$$v_p: C^1(\mathbb{E}^3, \mathbb{R}) \longrightarrow \mathbb{R}$$

$$f \longmapsto v_p[f] = \frac{d}{dt}f(p + t\, v_p)\Big|_{t=0}$$

which is the directional derivative of the function. This second version is crucial for generalising the concept of a vector tangent to a surface of arbitrary dimension independently of $\mathbb{R}^n$.

We can associate a unit vector U_p orthogonal to the tangent plane $T_p(M)$ at each point $p \in M$. The assignment

$$U: M \longrightarrow \mathbb{E}^3$$

$$p \longmapsto U_p$$

gives us, then, a unit vector field that is orthogonal to M at all points. U defines an 'orientation' for M, and we will assume that M is sufficiently regular for U to be well defined *locally,* and to be continuously differentiable (in the sense which we will define below).

The concepts of *tangent vector, tangent space,* and *orientation* will be defined formally after we introduce the concept of 'differentiable manifold' (see chapter 4). For the moment, we need no more elements than the ones defined above. From the vector field U we will build the form operator S, and from the latter we will obtain the normal curvature of M in any direction, its principal curvatures, and its Gaussian curvature. For this we will need to generalise the concept of the directional derivative which we have for functions to include vector fields. This generalisation is called the *covariant derivative.*

2.19 Definition (covariant derivative)

Let W be a vector field in $\mathbb{E}^3$, and let v be a vector tangent to $\mathbb{E}^3$ at p. Then

$$\nabla_v W \stackrel{\text{def}}{=} W'(p + t\, v)\big|_{t=0}$$

is the 'covariant derivative of W with respect to v at point p'. ($\nabla_v W$ measures the initial variation speed of $W(p)$ as p moves in the direction of v; see figure 2.8.)

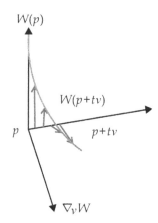

Figure 2.8. Construction of the covariant derivative of a vector field.

2.20 Example

Suppose that $W = x^2 \hat{e}_1 + y z \hat{e}_2$ and that $v = (-1, 0, 2)$ at $p = (2, 1, 0)$. Then $p + t v = (2 - t, 1, 2 t)$, so $W(p + t v) = (2 - t)^2 \hat{e}_1 + 2 t \hat{e}_2$. In other words,

$$\nabla_v W = W'(p + t v)|_{t=0} = -4 \hat{e}_1 + 2 \hat{e}_2$$

where, strictly speaking, $\hat{e}_1$ and $\hat{e}_2$ are evaluated at p.

2.21 Lemma (calculation of covariant derivatives)

Let $W = \omega^i \hat{e}_i$ be a vector field in $\mathbb{E}^3$, where we have used the repeated-index summation convention, and let $v \in T_p(\mathbb{E}^3)$ be a tangent vector to $\mathbb{E}^3$ at p. Then

$$\nabla_v W = v[\omega^i] \hat{e}_i(p) \qquad (2.1)$$

(i.e. in order to apply ∇_v to a vector field W, we apply v (as a differential operator) to its Euclidean coordinates (which are real-valued functions); the covariant derivative then inherits all the properties of the directional derivative).

Proof. $W(p + t v) = \omega^i(p + t v)\hat{e}_i(p + t v)$ is the restriction of W to the curve $t \mapsto p + t v$. Differentiating such a field at $t = 0$ amounts to simply differentiating its coordinates at $t = 0$: $\omega^i(p + t v)'|_{t=0}$, but this is precisely the definition of $v[\omega^i]$. Thus

$$\nabla_v W = W'(p + t v)|_{t=0} = v[\omega^i] \hat{e}_i|_p \qquad (2.2)$$

$\square$

2.22 Example

We take the vector field of example 2.21 $W = x^2 \hat{e}_1 + y z \hat{e}_2$, with $v = (-1, 0, 2)$ at $p = (2, 1, 0)$. We want to calculate $\nabla_v W = v[\omega^i] \hat{e}_i(p)$.

We have $\nabla x^2 = (2x, 0, 0)$ and thus $v[\omega^1] = v \cdot \nabla x^2 = -2x$. Similarly, $\nabla(y\,z) = (0, z, y)$ and $v[\omega^2] = v \cdot \nabla(y\,z) = 2\,y$. Then

$$\nabla_v W|_p = v[\omega^i]\,\hat{e}_i|_p = (-2\,x\,\hat{e}_1 + 2\,y\,\hat{e}_2)|_p = -4\,\hat{e}_1(p) + 2\,\hat{e}_2(p) \qquad (2.3)$$

since, at p, we have $x = 2$, $y = 1$, $z = 0$.

2.23 Definition (form operator)

Let M be a surface in $\mathbb{E}^3$ with an orientation given by the field U. The linear operator

$$S_p: T_p(M) \longrightarrow T_p(M)$$
$$v \longmapsto -\nabla_v U$$

is called the 'form operator' of M at p.

($T_p(M)$ consists of all the Euclidean vectors orthogonal to $U(p)$. Therefore, the rate of variation $\nabla_v U$ of U in the direction of v tells us how the planes tangent to M vary in the direction of v; this is an infinitesimal description of the way in which M itself curves in $\mathbb{E}^3$. A diagramatic sketch is shown in figure 2.9.)

2.24 Example

(i) Let M be a sphere of radius r, and let U be the unit vector orthogonal to M pointing 'outwards': $U = \frac{1}{r}\sum x^i\,\hat{e}_i$. Then

$$\nabla_v U = \frac{1}{r}\sum v[x^i]\,\hat{e}_i(p) = \frac{v}{r}$$

That is, $S(v) = -\frac{v}{r}$. The form operator S simply multiplies by the scalar $-\frac{1}{r}$: the sphere bends in the same way in all directions at all points.

(ii) Let M be a plane within $\mathbb{E}^3$. U is then parallel to itself at all points (it has constant Euclidean coordinates at all points), therefore

$$S(v) = -\nabla_v U = 0$$

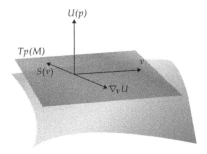

Figure 2.9. The form operator measures how the tangent plane to a surface varies as we move in a given direction.

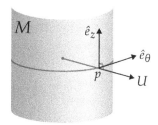

Figure 2.10. The form operator for a cylinder vanishes in the z-direction, and is equal to $-\frac{1}{r}$ in the direction of θ.

(iii) Let M be the cylinder in $\mathbb{E}^3$ given by $x^2 + y^2 = r^2$ (see figure 2.10). At each point $p \in M$, let $\hat{e}_z$ be the unit vector parallel to the cylinder's axis, $\hat{e}_\theta$ be the unit vector tangent to the circumference of a cross section at p, and $U = -\hat{e}_z \times \hat{e}_\theta$.

Then, if U moves along $\hat{e}_z$ it remains parallel to itself, i.e. $S(\hat{e}_z) = 0$, while if it moves along $\hat{e}_\theta$ it rotates in the same way as when it moves along the surface of a sphere of radius r: $S(\hat{e}_\theta) = -\frac{1}{r}\hat{e}_\theta$.

(iv) Let M be the saddle given by $z = x\,y$, and let $p = (0,0,0)$. The x and y axes are contained within M, and therefore $\hat{e}_1 = (1, 0, 0)$ and $\hat{e}_2 = (0, 1, 0)$ are tangent to M at p. In general, for

$$M: z = f(x, y)$$

with

$$f(0, 0) = f_x(0, 0) = f_y(0, 0) = 0$$

where

$$f_x = \frac{\partial f}{\partial x}, \quad f_y = \frac{\partial f}{\partial y}, \quad f_z = \frac{\partial f}{\partial z}$$

$\hat{e}_1(0)$ and $\hat{e}_2(0)$ are tangent to M at the origin and we have

$$U = \frac{-f_x\,\hat{e}_1 - f_y\,\hat{e}_2 + \hat{e}_3}{\sqrt{1 + f_x{}^2 + f_y{}^2}}$$

In our particular case,

$$U = \frac{-y\,\hat{e}_1 - x\,\hat{e}_2 + \hat{e}_3}{\sqrt{1 + x^2 + y^2}}$$

$$S(\hat{e}_1) = -\nabla_{\hat{e}_1} U = \frac{\hat{e}_2}{\sqrt{1 + x^2 + y^2}}$$

$$S(\hat{e}_2) = -\nabla_{\hat{e}_2} U = \frac{\hat{e}_1}{\sqrt{1 + x^2 + y^2}}$$

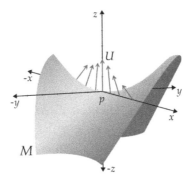

Figure 2.11. The shape of the form operator for a hyperbolic paraboloid.

Therefore, in general, $S(a\,\hat{e}_1 + b\,\hat{e}_2) = \frac{1}{\sqrt{1+x^2+y^2}}(b\,\hat{e}_1 + a\,\hat{e}_2)$. (Besides, at p = (0,0,0)

we have $\frac{1}{\sqrt{1+x^2+y^2}} = 1$.) Note that, if U moves in the direction of $\hat{e}_2$ (the y-axis), it

rotates in the direction of $-\hat{e}_1$ (the $-$x-axis), and vice versa, as shown in figure 2.11.

2.25 Lemma (geometry determines dynamics)

Let $M \subset \mathbb{E}^3$ be a surface in $\mathbb{E}^3$, $\alpha: I \longrightarrow M$ be a curve on M, and U be an orientation on M. Then

$$\alpha'' \cdot U = S(\alpha') \cdot \alpha'$$

Proof. Since α is a curve on M, $\alpha' \in T(M) = \bigcup_{p \in M} T_pM$. Therefore, $\alpha \cdot U = 0$. Differentiating,

$$\alpha'' \cdot U + \alpha' \cdot U' = 0$$

which is to say that

$$\alpha'' \cdot U = -U' \cdot \alpha' = S(\alpha') \cdot \alpha'$$

$\square$

($\alpha'' \cdot U$ is the normal component of the acceleration α'' of α. The lemma tells us that this component depends only on the velocity α' and on the form operator S. Therefore, all curves on M with a given velocity v at a point p have the same normal component of the acceleration at p: $S(v) \cdot v$. This is the acceleration component they must have because of the way in which M is curved. Here we can see a first example of how geometry determines the dynamics.)

2.26 Definition (normal curvature)

Let u be a unit vector tangent to $M \subset \mathbb{E}^3$ at p. Then

$$k(u) \overset{\text{def}}{=} S(u) \cdot u$$

is called 'normal curvature of M in the direction of u'.

2.27 Definition (principal curvatures)

$$k_1 = \max\{k(u)|u \in T_p(M)\}$$
$$k_2 = \min\{k(u)|u \in T_p(M)\}$$

are called 'principal curvatures of M at p'. The directions in which these occur are called 'principal directions of M at p'. If $k_1 = k_2$ (i.e. $k(u) = $ const.), p is called an 'umbilical point at M'.

2.28 Theorem (on umbilical points)

(i) Let p be an umbilical point of $M \subset \mathbb{E}^3$. Then, at p, S is merely the multiplication by the scalar $k = k_1 = k_2$.

(ii) Let p be a non-umbilical point of $M \subset \mathbb{E}^3$ (i.e. $k_1 \neq k_2$). Then there are exactly two principal directions and they are perpendicular to each other. Furthermore, if e_1 and e_2 are vectors in the principal directions then

$$S(e_1) = k_1 e_1, \qquad S(e_2) = k_2 e_2$$

(that is, k_1 and k_2 are eigenvalues of S, and e_1 and e_2 are its corresponding eigenvectors).

Proof. Suppose k's maximum value k_1 occurs in the direction of e_1:

$$k_1 = k(e_1) = S(e_1) \cdot e_1$$

Let e_2 be a unit vector tangent to M and perpendicular to e_1. Then there exists $\theta \in \mathbb{R}$ such that for all $u \in T_p(M)$ we have

$$u = u(\theta) = \cos(\theta)e_1 + \sin(\theta)e_2$$

and we can see k as a function on $\mathbb{R}$: $k(u(\theta)) = k(\theta)$. Then

$$\begin{aligned}
k(\theta) &= S(\cos(\theta)e_1 + \sin(\theta)e_2) \cdot (\cos(\theta)e_1 + \sin(\theta)e_2) \\
&= \cos^2(\theta)S_{11} + 2\cos(\theta)\sin(\theta)S_{12} + \sin^2(\theta)S_{22}
\end{aligned} \tag{2.4}$$

where we have set $S_{ij} \equiv S(e_i) \cdot e_j$ and used the symmetry $S_{12} = S_{21}$, which we shall prove below. Differentiating,

$$\frac{dk(\theta)}{d\theta} = 2\cos(\theta)\sin(\theta)(S_{22} - S_{11}) + 2(\cos^2(\theta) - \sin^2(\theta))S_{12} \tag{2.5}$$

Since, by the hypothesis, k takes its maximum value in the direction of e_1 ($\theta = 0$),

$$\frac{dk}{d\theta}\bigg|_{\theta=0} = 0$$

Therefore, by equation (2.5),

$$S_{12} = 0$$

Hence, from equation (2.4),

$$S(e_1) = S_{11}\, e_1, \qquad S(e_2) = k_2\, e_2 \tag{2.6}$$

Case a: p is umbilical: $S_{22} = k(e_2) = k(e_1) = S_{11} = k_1$. Then, by equation (2.6), S is merely the multiplication by k_1.

Case b: p is not umbilical. From equation (2.4),

$$k(\theta) = \cos^2(\theta)k_1 + \sin^2(\theta)S_{22} \tag{2.7}$$

k_1 is the maximum value taken by $k(\theta)$, and $k(\theta) \neq$ const. Therefore,

$$k_1 > S_{22}$$

From equation (2.7) we can see that k_1 (the maximum value) is only reached by $k(\theta)$ when $\cos\theta = \pm 1$ and $\sin\theta = 0$ (i.e. in the direction of e_1) and that the minimum value of $k(\theta)$ is S_{22}, which is reached when $\cos\theta = 0$ and $\sin\theta = \pm 1$ (i.e. in the direction of e_2). Therefore, by equation (2.6),

$$S(e_1) = k_1\, e_1, \qquad S(e_2) = k_2\, e_2$$

We now need to prove that $S_{12} = S_{21}$, as assumed above. To do so, observe that $U \cdot e_1 = 0$. Differentiating covariantly with respect to e_2,

$$0 = (\nabla_{e_2} U) \cdot e_1 + U \cdot (\nabla_{e_2} e_1) = -S(e_2) \cdot e_1 + U \cdot \sum_i e_2[e_1^i]$$

$$= -S_{21} + U \cdot \sum_{i,j} e_2^j \left(\frac{\partial}{\partial x^j} e_1^i\right)\frac{\partial}{\partial x^i}$$

and similarly for S_{12} by interchanging the indices $1 \leftrightarrow 2$. Therefore,

$$S_{12} - S_{21} = U \cdot \left[\sum_{i,j} e_1^i\left(\frac{\partial}{\partial x^i}e_2^j\right)\frac{\partial}{\partial x^j} - \sum_{i,j} e_2^j\left(\frac{\partial}{\partial x^j}e_1^i\right)\frac{\partial}{\partial x^i}\right]$$

Recall that e_1 and e_2 are not necessarily in the directions of $\frac{\partial}{\partial x^1}$ and $\frac{\partial}{\partial x^2}$, respectively, since e_1 was taken in the direction in which S takes its maximum value and e_2 was taken tangent to the surface and perpendicular to e_1. Let u^1, u^2 be coordinates adapted to e_1, e_2; in other words, $e_1 = \frac{\partial}{\partial u^1}$, $e_2 = \frac{\partial}{\partial u^2}$. In these coordinates,

$$e_{1\{u\}}^i = \delta_1^i, \qquad e_{2\{u\}}^j = \delta_2^j$$

and, as we shall see in chapter 6,

$$e_1^i = A_k^i e_{1\{u\}}^k = \frac{\partial x^i}{\partial u^k} \delta_1^k = \frac{\partial x^i}{\partial u^1}$$

$$e_2^j = A_k^j e_{2\{u\}}^k = \frac{\partial x^j}{\partial u^k} \delta_2^k = \frac{\partial x^j}{\partial u^2}$$

Therefore,

$$S_{12} - S_{21} = U \cdot \left[\sum_{i,j} \frac{\partial x^i}{\partial u^1} \frac{\partial}{\partial x^i} \frac{\partial x^j}{\partial u^2} \frac{\partial}{\partial x^j} - \sum_{i,j} \frac{\partial x^j}{\partial u^2} \frac{\partial}{\partial x^j} \frac{\partial x^i}{\partial u^1} \frac{\partial}{\partial x^i} \right]$$

$$= U \cdot \left[\frac{\partial^2}{\partial u^1 \partial u^2} - \frac{\partial^2}{\partial u^2 \partial u^1} \right] = 0$$

$\square$

2.29 Corollary (Euler's formula)

Let k_1, k_2 be the principal curvatures and e_1, e_2 the principal vectors of $M \subset \mathbb{E}^3$ at p. Let $u = \cos(\theta)e_1 + \sin(\theta)e_2$. Then

$$k(u) = k_1 \cos^2 \theta + k_2 \sin^2 \theta$$

(that is, the principal curvatures control the shape of M in the vicinity of any given point).

Proof. The proof follows trivially from the equations in the proof of theorem 2.28.

$\square$

2.30 Definition (Gaussian and mean curvatures)

$K = \det(S) = \det(S_{ij})$ is called the 'Gaussian curvature of M'.

$H = \frac{1}{2} \operatorname{tr}(S)$ is called 'mean curvature of M'.

(Both are real-valued functions on M.)

2.31 Lemma (expressions for Gaussian and mean curvatures)

$$K = k_1 k_2, \qquad H = \frac{k_1 + k_2}{2}$$

Proof. The matrix of S with respect to the principal directions e_1 and e_2 is

$$S = \begin{pmatrix} k_1(p) & 0 \\ 0 & k_2(p) \end{pmatrix}$$

because $S(e_1) = k_1 e_1$ and $S(e_2) = k_2 e_2$. Therefore,

$$K = \det(S) = k_1 k_2, \qquad H = \frac{1}{2} \, \text{tr}(S) = \frac{k_1 + k_2}{2}$$

Since the trace and the determinant are invariant under coordinate transformations, the result follows. □

2.32 Remark

(i) If we change our choice of orthogonal unit vector field from U to $-U$, both k_1 and k_2 change sign. Therefore, K is independent of such a choice, while H is not.

(ii) If $K(p) > 0$, then $k_1(p)$ and $k_2(p)$ have the same sign. Therefore, by corollary 2.29, either $k(u) > 0 \; \forall u \in T_p(M)$ or $k(u) < 0 \; \forall u \in T_p(M)$. M then bends in such a way that it recedes from its tangent plane $T_p(M)$ towards a single side in all directions (that side may be called 'upwards' or 'downwards')—in other words, M is shaped, locally, like an ellipsoid (see figure 2.12).

(iii) If $K(p) < 0$, then $k_1(p)$ and $k_2(p)$ have opposite signs. Therefore, M bends in such a way that it recedes from $T_p(M)$ in one direction along e_1 and in the opposite direction along e_2—in other words, M is shaped, locally, like a saddle (see figure 2.13, left).

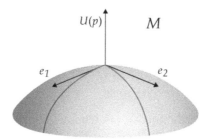

Figure 2.12. Local shape of the surface when $K(p) > 0$.

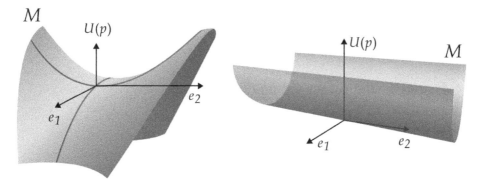

Figure 2.13. Left: local shape of the surface when $K(p) < 0$. Right: local shape of the surface when $K(p) = 0$, for the cases $k_1(p) \neq 0$ and $k_2(p) = 0$, or $k_1(p) = 0$ and $k_2(p) \neq 0$.

(iv) If $K(p) = 0$, one of several things may occur. If $k_1(p) \neq 0$ and $k_2(p) = 0$, or if $k_1(p) = 0$ and $k_2(p) \neq 0$, then M is shaped locally like a cylinder. If, on the other hand, $k_1(p) = k_2(p) = 0$, then M is shaped locally like a plane (see figure 2.13, right).

2.33 Calculation of curvatures

Let $x \colon D \subset \mathbb{R}^2 \longrightarrow M \subset \mathbb{E}^3$, $x(u, v) = (x^1(u, v), x^2(u, v), x^3(u, v))$ be a parametrisation of a region of M in terms of coordinates. We use the notation $x_u = \frac{\partial x}{\partial u}$, $x_v = \frac{\partial x}{\partial v}$ and define

$$E := x_u \cdot x_u, \qquad F := x_u \cdot x_v, \qquad G := x_v \cdot x_v$$

E and G are, respectively, the squares of the speed of the curves parametrised by u and v; F is a measure of the angle between said curves. Put differently, E, F and G are the 'distortion functions' that measure the way in which x distorts region D of the plane to turn it into the curved region $x(D)$.

$x_u \times x_v$ is perpendicular to both x_u and x_v, so it is perpendicular to M at $x(u, v)$. Therefore,

$$U \overset{\text{def}}{=} \frac{x_u \times x_v}{\|x_u \times x_v\|}$$

has unit magnitude and is orthogonal to M. U then defines a local orientation on M. (x_u and x_v are not orthogonal to each other, so $\{x_u, x_v, U\}$ is not an orthonormal reference system.)

Let S be the form operator associated to U.

We define

$$l := S(x_u) \cdot x_u, \qquad m := S(x_u) \cdot x_v = S(x_v) \cdot x_u, \qquad n := S(x_v) \cdot x_v$$

Since $\{x_u, x_v, U\}$ is not an orthonormal reference system, l, m, and n might not have simple expressions in terms of x_u and x_v, but they do provide us with simple expressions for the curvatures:

$$K(x) = \frac{l\,n - m^2}{E\,G - F^2}, \qquad H(x) = \frac{G\,l + E\,n - 2\,F\,m}{2(E\,G - F^2)}$$

where $E\,G - F^2 = \|x_u \times x_v\|^2$.

2.34 Example (the torus, yet again!)

For the torus $M = \mathbb{T}^2$, given by

$$x(u, v) = (h(u)\cos v, h(u)\sin v, g(u))$$

with $h(u) = R + r\cos u$, $g(u) = r\sin u$, we have

Figure 2.14. Curvature on different regions of the torus.

$$E = r^2, \qquad F = 0, \qquad G = (R + r \cos u)^2$$
$$l = r, \qquad m = 0, \qquad n = (R + r \cos u)\cos u$$

so that

$$K = \frac{\cos u}{r(R + r \cos u)}$$

and (see figure 2.14)

$K > 0$ on the outer half of the torus (O)

$K < 0$ on the inner half of the torus (J)

$K = 0$ on the top and bottom circumferences ($u = \pm\pi/2$) (L)

$K_{max} = \dfrac{1}{r(R + r)}$ occurs on the outer equator ($u = 0$)

$K_{min} = \dfrac{-1}{r(R - r)}$ occurs on the inner equator ($u = \pi$)

Chapter 3

Elements of topology

'Topology' is the branch of mathematics that studies continuity. The mental picture is that of continuous deformations of one object into another. For example, a solid triangle may be continuously deformed into a solid square, or into a disc, and so we say that they are 'topologically equivalent', but a triangle with a hole inside is *not* topologically equivalent to a solid square or disc, since one cannot create or destroy holes through continuous transformations.

Having said that, topology can also be used to study discrete sets, and the set's structure is determined by the set's topology, that is, any given set will have different structures if we define different topologies on it, as we shall soon see.

We denote by $\mathcal{P}(M) = 2^M$ the 'power of M', the set of all subsets of M. Given $V \subset M$, we denote by $V^c = M/V$ the complement of V in M. In this chapter we will define many concepts in topology which will allow us to do calculus on manifolds, which we build upon throughout the following chapters.

3.1 Definition (topological space)

Let M be a set, and let $\tau \subset \mathcal{P}(M) = 2^M$. Then (M, τ) is called a 'topological space' if and only if:

(i) $\varnothing \in \tau$, $M \in \tau$;

(ii) $U_1, \ldots, U_n \in \tau \ \Rightarrow \ \bigcap_{i=1}^{n} U_i \in \tau$;

(iii) $\{U_\alpha\}_{\alpha \in A} \subset \tau \ \Rightarrow \ \bigcup_{\alpha \in A} U_\alpha \in \tau$ (with A a set of arbitrary indices).

3.2 Definition (nomenclature)

Let (M, τ) be a topological space. Then

(i) τ is called a 'topology on M';

(ii) $U \in \tau$ is called an 'open set within M';

(iii) $V \subset M$ is called a 'closed set within M' if and only if $V^c \in \tau$;

(iv) if $\tau = \mathcal{P}(M) = 2^M$ then τ is called the 'discrete topology' of M;

(v) if $x \in U \in \tau$ then U is called a 'neighbourhood of x'.

3.3 Definition (connected space)

A topological space (M, τ) is called 'connected' if there are no two open, non-empty sets U, V such that $U \cap V = \emptyset$ and $U \cup V = M$.

3.4 Lemma (connectedness)

For any given topological space (M, τ), the following statements are equivalent:

(i) (M, τ) is connected;

(ii) if $A \subset M$ is both open and closed, then $A = \emptyset$ or $A = M$.

Proof. To prove that (i) $\Rightarrow$ (ii), let A be an open and closed set within M such that $A \neq \emptyset$ and $A \neq M$. Since A is open and closed, A^c is closed and open. Now, $A \cup A^c = M$ and $A \cap A^c = \emptyset$, and both A and A^c are open. Therefore, M is disconnected. (!)

To prove that (ii) $\Rightarrow$ (i), suppose M is not connected. Then $\exists\, B_1, B_2 \in \tau$ such that $B_1 \cup B_2 = M$, $B_1 \cap B_2 = \emptyset$, $B_1 \neq \emptyset \neq B_2$. Then $B_1{}^c = B_2$ and, as $B_1 \in \tau$ (B_1 is open), B_2 is closed. Therefore, $B_2 \subset M$ is both open and closed. But $B_1 \neq \emptyset \neq B_2$, so $B_2 \neq \emptyset$ and $B_2 \neq M$. (!) $\qquad\square$

The following definitions, called *separability axioms*, are a means of making sure our space has 'enough' open sets and they are well organised.

3.5 Definition (T_0)

(M, τ) is a 'T_0' or 'Komolgorov' topological space if and only if $\forall\, x, y \in M$ such that $x \neq y$ and there exists $U \in \tau$ such that $x \in U$, $y \notin U$. In other words, (M, τ) is a 'T_0' or 'Komolgorov' topological space if and only if, given two different points of M, they are topologically distinguishable (by means of an open set).

3.6 Example

Let $M = \{1, 2, 3, 4\}$, $\tau_1 = \{\emptyset, \{1, 2\}, \{1, 2, 3\}, \{1, 2, 4\}, M\}$, $\tau_2 = \{\emptyset, \{1\}, \{1, 2\}, \{1, 2, 3\}, M\}$. Then (M, τ_1) is not T_0 but (M, τ_2) is.

3.7 Definition (T_1)

(M, τ) is a 'T_1' or 'Frèchet' topological space if and only if $\forall\, x, y \in M$ such that $x \neq y$ and there exist $U, V \in \tau$ such that $x \in U$, $y \in V$, $y \notin U$, $x \notin V$. In other words, (M, τ) is a 'T_1' or 'Frèchet' topological space if and only if points are closed sets within M.

3.8 Example

Let M be a finite set and (M, τ) be a T_1 topological space. Let $p, q \in M$. Finally, let B_p be the smallest open set containing p (i.e. $B_p = \bigcap \{U_i \in \tau$ such that $p \in U_i)$ and let B_q be the smallest open set containing q. Since M is T_1, $p \notin B_q$ and $q \notin B_p$. Therefore, M is totally disconnected, that is, τ is the *discrete topology* of M.

3.9 Definition (T_2)

(M, τ) is a 'T_2' or 'Hausdorff' topological space if and only if $\forall x, y \in M$ such that $x \neq y$ there exist $U, V \in \tau$ such that $x \in U$, $y \in V$, $U \cap V = \varnothing$.

3.10 Example

(i) Let $M = \{x_1, x_2\}$, $\tau = \{\varnothing, \{x_1\}, M\}$. Then (M, τ) is not a Hausdorff topological space.

(ii) Consider the set M formed by $\mathbb{R}$ with two copies of the nonnegative numbers, and let τ be the usual topology of $\mathbb{R}$ (i.e. τ contains the usual open sets of $\mathbb{R}$, as shown in figure 3.1):

(M, τ) is not a Hausdorff topological space because the two zeros have no nonintersecting open neighbourhoods.

(iii) Let $M = \mathbb{R} \cup \{p\}$ with the usual open sets of $\mathbb{R}$ and with neighbourhoods of p given by $U\backslash\{0\} \cup \{p\}$ such that U is a neighbourhood of $0 \in \mathbb{R}$ (see figure 3.2):

$p \neq 0$, yet there are no nonintersecting open neighbourhoods of p and 0; therefore, (M, τ) is not a T_2 topological space.

Definitions 3.7 and 3.9 are two of the most important definitions in topology: we can separate points by means of open sets. This is already beginning to look like the spaces we know.

There are more axioms of separability; these focus on the separability of closed sets. They are $T_{\frac{1}{2}}$, $T_{2\frac{1}{2}}$, T_3, $T_{3\frac{1}{2}}$, T_4, T_5 and T_6.

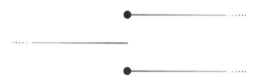

Figure 3.1. An example of a set which is *not* Hausdorff.

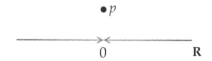

Figure 3.2. Yet another example of a set which is *not* Hausdorff.

The following two definitions bound the number of 'sufficient' open sets.

3.11 Definition (basis)

Let M be a set. A 'basis' of M is a collection $\mathcal{B}$ of subsets of M such that
(i) $\forall x \in M \; \exists B \in \mathcal{B}$ such that $x \in B$—or, in other words,

$$M = \bigcup_{B \in \mathcal{B}} B$$

(ii) if $B_1, B_2 \in B$ and $x \in B_1 \cap B_2$ then $\exists B_3 \in \mathcal{B}$ such that $x \in B_3 \subset B_1 \cap B_2$.

3.12 Definition (second-countable topological space)

A topological space (M, τ) is 'second-countable' if it allows a countable basis.

The concept of 'first-countable' topological space exists, but it is weaker than that of 'second-countable' topological space. Given $x \in M$, the collection $\mathcal{B}_x$ of neighbourhoods of x is called a 'local basis at x' if every neighbourhood of x contains a subset $B \in \mathcal{B}_x$. A topological space is 'first-countable' if there is a countable local basis $\forall \, x \in M$.

3.13 Definition (continuous function)

Let (M, τ_1) and (N, τ_2) be topological spaces. A function $f : M \longrightarrow N$ is called 'continuous' if and only if $\forall \, V \in \tau_2$ we have $f^{-1}(V) \in \tau_1$ (i.e. the inverse image of an open set is open).

For metric spaces, this definition is equivalent to the definition in terms of ϵ and δ of elementary calculus.

3.14 Definition (homeomorphism)

Let (M, τ_1) and (N, τ_2) be topological spaces, and let $U \in \tau_1$, $V \in \tau_2$. Finally, let $\varphi : U \longrightarrow V$ be continuous. φ is called a 'homeomorphism from U to V' if and only if $\varphi^{-1} : V \longrightarrow U$ exists and is continuous.

In particular, if $U = M$ and $V = N$ and φ is a homeomorphism from M to N, then M and N are called 'homeomorphic'.

The fact that for a homeomorphism φ, both φ and φ^{-1} are continuous, implies that not only does φ map points of M to N bijectively, but it also maps open sets of M to open sets of N bijectively. This means that M and N are topologically equivalent and that any topological property of M also holds for N (and vice versa).

The concept of homeomorphism automatically gives us the idea of 'topological invariants', that is, properties which do not change under homeomorphisms. These may be simple numbers such as the dimension n of $\mathbb{R}^n$, topological concepts such as compactness and connectedness, or mathematical structures such as homotopy groups, homology groups, and cohomology groups. The study of these structures gives rise to *algebraic topology*, a branch of mathematics which reduces topological problems to algebraic ones in order to take advantage of all the tools of algebra for their solution.

In what follows we shall see some topological invariants. A brief look into the concepts of homology and cohomology will be delayed until chapter 10.

3.15 Theorem (compactness as topological invariant)

Let M_1, M_2 be topological spaces, with M_1 compact.
Let $\varphi : M_1 \to M_2$ be a homeomorphism between M_1 and M_2.
Then M_2 is compact.

Proof. Let $\{U_\alpha\}_{\alpha \in I}$ be an open cover of M_2. Since φ is continuous, $\varphi^{-1}(U_\alpha)$ is open in M_1 for all α; and since φ is invertible, $\bigcup_{\alpha \in I} \{\varphi^{-1}(U_\alpha)\}$ is an open cover of M_1.

But M_1 is compact, so there exist $\alpha_{i_1}, \dots, \alpha_{i_n} \in I$ such that $\varphi^{-1}(U_{\alpha_{i_1}}) \bigcup \cdots \bigcup \varphi^{-1}(U_{\alpha_{i_n}})$ is a finite open cover of M_1.

Therefore $U_{\alpha_{i_1}} \bigcup \cdots \bigcup U_{\alpha_{i_n}}$ is an open finite subcover of M_2, i.e. M_2 is compact. $\square$

3.16 Theorem (connectedness as a topological invariant)

Let M_1, M_2 be topological spaces, with M_1 connected.
Let $\varphi : M_1 \to M_2$ be a homeomorphism between M_1 and M_2.
Then M_2 is connected.

Proof. Let us suppose that M_2 is not connected, and express it as $M_2 = M_2{}^A \bigcup M_2{}^B$ with $M_2{}^A$, $M_2{}^B$ open and $M_2{}^A \bigcap M_2{}^B = \varnothing$.

The fact that φ is continuous implies that $\varphi^{-1}(M_2{}^A)$ y $\varphi^{-1}(M_2{}^B)$ are open and disjoint. But then $\varphi^{-1}(M_2{}^A) \bigcup \varphi^{-1}(M_2{}^B) = M_1$, which would be a contradiction. Therefore, M_2 must be connected. $\square$

3.17 Theorem (dimension as a topological invariant)

The dimension of a topological space is invariant under homeomorphisms.

Proof. We will carry out the proof for the spaces $\mathbb{R}^n$.
   (i) Suppose that there exists a homeomorphism $\varphi : \mathbb{R} \to \mathbb{R}^2$. Let us think of $\mathbb{R}^2$ as the x,y plane and of $\mathbb{R}$ as the x-axis, and let us remove the origin $x = 0$ from $\mathbb{R}$ and the point $\varphi(0)$ from $\mathbb{R}^2$.

   Then $\mathbb{R}\backslash\{0\}$ is disconnected but $\mathbb{R}^2\backslash\{\varphi(0)\}$ is connected, while $\varphi|_{\mathbb{R}\backslash\{0\}}$ is a homeomorphism between the two, which would contradict theorem 3.16.
   (ii) Let us now suppose that we can build a homeomorphism $\varphi : \mathbb{R}^2 \to \mathbb{R}^3$. Then $\varphi : \mathbb{R}^2\backslash\{(0, 0)\} \to \mathbb{R}^3\backslash\{\varphi(0, 0)\}$ would be a homeomorphism too.
   Consider the circle $x^2 + y^2 = a^2$, and the limit $a \to \infty$.

In $\mathbb{R}^3\backslash\{\varphi(0, 0)\}$ we can deform the circle down to a point since, if necessary, we deform it away from the x,y plane. However, in $\mathbb{R}^2\backslash\{(0, 0)\}$ we cannot deform the circle down to a point.

This would break the continuity of φ when restricted to the circle, thus the homeomorphism cannot exist.

(iii) By induction we may prove, in the same manner as in (ii), that

$$\mathbb{R}^n \text{ homeomorphic to } \mathbb{R}^m \Longleftrightarrow n = m$$

by surrounding the origin of $\mathbb{R}^n$ with the $(n - 1)$-sphere S^{n-1}.

$\square$

3.18 The fixed-point problem (Brower's theorem)

Let $f: M \to M$ be a function from a topological space M unto itself. We can ask the question, does f have any fixed points? We sketch a proof for $M = \mathbb{R}^n$:

(i) $\mathbb{R}^1$ $f: (a, b) \to (a, b)$ and the question reduces itself to whether the graph of f crosses or does not cross the line $f(x) = x$ (see figure 3.3), and when the domain is bounded it necessarily crosses it.

(ii) $\mathbb{R}^2$ Let B^2 be the two-dimensional disk defined by $x^2 + y^2 \leqslant a^2$, and take B^2 as the domain of f. Let v be a vector defined at every point of B^2 by $v(x) = f(x) - x$, and note the following:

if x lies in the boundary of B^2, as shown in figure 3.4, and we rotate it once by 2π about the centre of B^2, then $v(x)$ will also rotate once by 2π. We say that '$v(x)$ has index 1'. The 'index of a vector field' is the number of full turns that it suffers when we rotate the variable by 2π, and it happens to be a topological invariant. Therefore, under a continuous deformation of the border of B^2 (as into the dotted line in figure 3.5) the index of $v(x)$ will continue to be 1.

But if $v(x_0) = 0$ for some $x_0 \in B^2$, i.e. x_0 is a fixed point of f, then the index of v is not well defined at x_0.

Now suppose that $v(p) \neq 0$ for an arbitrary point p in the interior of B^2, and define

$$L = |v(p)| = |f(p) - p|$$

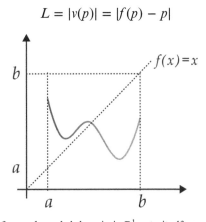

Figure 3.3. A function from a bounded domain in $\mathbb{R}^1$ onto itself necessarily has a fixed point.

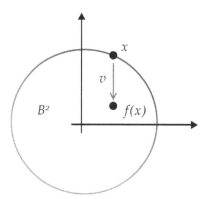

Figure 3.4. Vector field $v(x) = f(x) - x$ defined on the two-dimensional disk, in order to prove Brower's theorem.

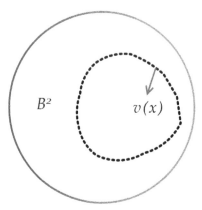

Figure 3.5. The index of a vector field does not change under a continuous deformation of the disk boundary.

Let C be the circle of radius $\epsilon \leqslant L$ about p, and let C' be its image under f. By continuity of f we may take ϵ sufficiently small such that $\bar{C} = \overset{\circ}{C} \cup \partial C$ and $\overline{C'} = \overset{\circ}{C'} \cup \partial C'$ do not intersect. (Here, $\overset{\circ}{C}$ denotes the interior and ∂C the boundary of C, and the same for C'.) This is possible since $f(\overset{\circ}{C}) = \overset{\circ}{C'}$ and $f(\partial C) = \partial C'$.

Then, as x is carried through a complete turn around C, $v(x)$ will not turn since the circles do not intersect, i.e. its index would be 0. However, we know that its index must be 1, so there must exist a point $p_0 \in B^2$ for which $v(p_0) = 0$, i.e. at which $f(p_0) = p_0$.

(iii) $\mathbb{R}^n$ The construction in (ii) may be extended to B^n for $n \in \mathbb{N}$.

3.19 Remark (importance of the fixed-point theorem)

Apart from being a surprising result (for instance, if we move around the surface of a liquid in a cup, without breaking it (as can be done using a magnetic stirrer), at least one point will remain in its starting place!), the result has important applications:

(i) For a differential operator A the existence of solutions to $A[f] = 0$ is equivalent to the existence of fixed points of the operator $A + I$ (with I the identity operator).

(ii) For a dynamical system

$$\frac{dx^i}{dt} = f^i(x^1, \ldots, x^n), \quad i = 1, \ldots, n$$

or, equivalently, $\dot{x} = f$, the singularities (zeroes) of the vector field $\dot{x}$ are the equilibrium configurations of the system, i.e. the critical points of f. Its study gives rise to *Morse's theory*.

Even though we shall only be working with metric spaces towards the end of this book, just for completeness we end this chapter with the definition of a distance function, and a lemma.

3.20 Definition (metric space)

Let (M, τ) be a topological space. Then:
  (i) $d : M \times M \longrightarrow \mathbb{R}$ is called a 'distance function over M' if and only if $\forall\, x, y \in M$ the following is true:
  (a) $d(x, y) \geqslant 0$.
  (b) if $d(x, y) = 0$ then $x = y$.
  (c) $d(x, z) \leqslant d(x, y) + d(y, z)$.
  (ii) (M, τ, d) is called a 'metric space'.

A function $d : M \times M \longrightarrow \mathbb{R}$ with the above properties is also called a 'metric over M'. In this book, the name 'metric' will be reserved for something more general which will be defined in chapter 15.

3.21 Lemma

Let M be an arbitrary set, and let d be a distance function over M. Then M has the structure of a topological space.

Proof. Trivial. We merely define $S_x(\epsilon) = \{y \in M \mid d(x, y) < \epsilon\}$ for $\epsilon \in \mathbb{R}^+$ and $x \in M$, and then we take $\{S_x(\epsilon)\}_{x \in M,\, \epsilon \in \mathbb{R}^+}$ as a basis for the topology of M. $\qquad \square$

Chapter 4

Differentiable manifolds

Differentiable manifolds are a generalisation of surfaces. Unlike the latter, however, we need not imagine a manifold as being immersed in a higher-dimensional space in order to study its geometric properties. In this chapter and the next, we define differentiable manifolds and build the basics to do calculus on them.

For simplicity, we shall henceforth denote a topological space by M, without referencing its topology τ. In addition, we will often call topological spaces simply 'spaces'.

4.1 Definition (locally Euclidean space)

A topological space M is called 'locally Euclidean' if and only if:
   (i) M is a second-countable Hausdorff space.
   (ii) $\exists\, n \in \mathbb{N}$ such that $\forall\, x \in M$ and there exist a vicinity $U \subset M$ of x, an open set $V \subset \mathbb{R}^n$ and a homeomorphism $\varphi_x \colon U \longrightarrow V$.

n is called the 'dimension of M' and is denoted as $\dim(M) = n$.

4.2 Examples

   (i) $\mathbb{R}^n$ is locally Euclidean: for any given $x \in \mathbb{R}^n$, take $\varphi_x(p) = p$.
   (ii) An open set $U \subset \mathbb{R}^n$ is locally Euclidean, which is shown trivially, as in (i), by taking the identity map.
   (iii) The n-dimensional unit sphere, $\mathbb{S}^n$, is locally Euclidean: for any given $x \in \mathbb{S}^n$, take $y \in \mathbb{S}^n$ such that $y \neq x$ and take φ_x to be the stereographic projection from y. The stereographic mapping $\varphi \colon \mathbb{S}^n \longrightarrow \mathbb{R}^n$ is given by projecting from the 'north pole' (y) of $\mathbb{S}^n$ to the hyperplane $\mathbb{R}^n$ which bisects $\mathbb{S}^n$ at the equator and then identifying the corresponding points that intersect $\mathbb{S}^n$ and $\mathbb{R}^n$ (see figures 4.1 and 4.2). For $\mathbb{S}^1$,

doi:10.1088/2053-2563/aadf65ch4

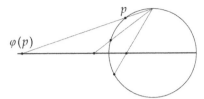

Figure 4.1. Stereographic projection for $\mathbb{S}^1$.

Figure 4.2. Stereographic projection for $\mathbb{S}^2$.

$$\varphi(x^1, x^2) = \frac{x^1}{1 - x^2} \in \mathbb{R}^n$$

where $(x^1)^2 + (x^2)^2 = 1$.

For $\mathbb{S}^2$,

$$\varphi(x^1, x^2, x^3) = \left(\frac{x^1}{1 - x^3}, \frac{x^2}{1 - x^3} \right) \in \mathbb{R}^2$$

where $(x^1)^2 + (x^2)^2 + (x^3)^2 = 1$.

In general, for $\mathbb{S}^n$,

$$\varphi(x^1, \ldots, x^{n+1}) = \left(\frac{x^1}{1 - x^{n+1}}, \ldots, \frac{x^n}{1 - x^{n+1}} \right) \in \mathbb{R}^n$$

where $(x^1)^2 + \ldots + (x^{n+1})^2 = 1$. Note that all points of $\mathbb{S}^n$ but the north pole are mapped to $\mathbb{R}^n$ (the north pole maps to the point at ∞) bijectively. In this sense, $\mathbb{S}^n$ is topologically like $\mathbb{R}^n \cup \{\infty\}$.

(iv) The n-dimensional projective space, $\mathbb{P}^n$, is locally Euclidean: since $\mathbb{S}^n$ covers $\mathbb{P}^n$ twice, for any given $x \in \mathbb{P}^n$ there exist open sets $\mathcal{N}_x \subset \mathbb{P}^n$ and $V \subset \mathbb{S}^n$ such that $x \in \mathcal{N}_x$ and $\mathcal{N}_x$ is homeomorphic to V ($\mathcal{N}_x \simeq V$). As we saw in (ii), V is homeomorphic to an open set $A \subset \mathbb{R}^n$. If $\mathcal{N}_x \simeq V$ by means of a homeomorphism φ_1 and $V \simeq A$ by means of a homeomorphism φ_2, then $\varphi_2 \circ \varphi_1$ is also a homeomorphism and $\mathcal{N}_x \simeq A$ by means of $\varphi_2 \circ \varphi_1$.

4.3 Exercise

Prove that the set of non-singular $n \times n$ matrices is locally Euclidean.

Solution. Identify each $n \times n$ matrix with the vector $v \in \mathbb{R}^{n^2}$ formed by writing the matrix's rows one after the other. The non-singular matrices then form an open set in $\mathbb{R}^{n^2}$: $\Delta^{-1}(\mathbb{R}\backslash\{0\})$, where $\Delta: \mathbb{R}^{n^2} \longrightarrow \mathbb{R}$ is the determinant function.

4.4 Definition (differentiable manifold)

Let M be an n-dimensional second-countable Hausdorff space, and let Φ be a collection of functions mapping open subsets of M to open subsets of $\mathbb{R}^n$ with the following properties:

    (i) $\{\mathrm{dom}(\varphi)|\ \varphi \in \Phi\}$ is an open cover of M.
    (ii) $\forall \varphi \in \Phi$, φ is a homeomorphism to an open set in $\mathbb{R}^n$.
    (iii) $\forall \varphi, \psi \in \Phi$ such that $\mathrm{dom}(\varphi) \cap \mathrm{dom}(\psi) \neq \varnothing$, we have

$$\psi \circ \varphi^{-1} \colon \varphi(\mathrm{dom}(\varphi) \cap \mathrm{dom}(\psi)) \longrightarrow \mathbb{R}^n \in C^k$$

(M, Φ) is then called an 'n-dimensional differentiable manifold of class C^k'. Furthermore, Φ is called an 'atlas' for M and each element φ of Φ is called a 'chart'. Φ is responsible for M's differentiable structure.

4.5 Definition (maximal atlas)

Let (M, Φ) be an n-dimensional differentiable manifold of class C^k. Then Φ is called a 'maximal atlas' if and only if the following is true: given any homeomorphism $\psi \colon U \subset M \longrightarrow \mathbb{R}^n$ with the property that for all $\varphi \in \Phi$ with $\mathrm{dom}(\varphi) \cap \mathrm{dom}(\psi) \neq \varnothing$ we have

    (i) $\psi \circ \varphi^{-1} \colon \varphi(\mathrm{dom}(\varphi) \cap \mathrm{dom}(\psi)) \longrightarrow \mathbb{R}^n \in C^k$,
    (ii) $\varphi^{-1} \circ \psi \colon \psi(\mathrm{dom}(\varphi) \cap \mathrm{dom}(\psi)) \longrightarrow \mathbb{R}^n \in C^k$,

then this homeomorphism ψ is an element of $\Psi \colon \psi \in \Phi$ (i.e. all compatible charts with those in Φ are already contained in Φ).

Note the following:

    (i) All differentiable manifolds of class C^k are locally Euclidean.
    (ii) Any locally Euclidean space gives rise to a differentiable manifold of class C^0.
    (iii) In addition to being called a 'chart' on M, $\varphi \in \Phi$ is called a 'local coordinate system'.

4.6 Exercise

Is a Möbius strip (see figure 4.3) a differentiable manifold?

Solution. Yes. Φ consists of only two charts, which are given by

$$\varphi^{-1}(u, v) = \left(\cos u + v \cos \frac{u}{2} \cos u,\ \sin u + v \cos \frac{u}{2} \sin u,\ v \sin \frac{u}{2} \right)$$

with

$$u \in (0, 2\pi), \quad v \in \left[-\frac{1}{2}, \frac{1}{2} \right]; \qquad u \in (-\pi, \pi), \quad v \in \left[-\frac{1}{2}, \frac{1}{2} \right]$$

each of which is a diffeomorphism (i.e. an infinitely differentiable homeomorphism).

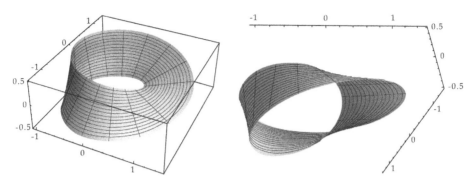

Figure 4.3. The Möbius strip is a differentiable manifold. See the text for the construction of the charts.

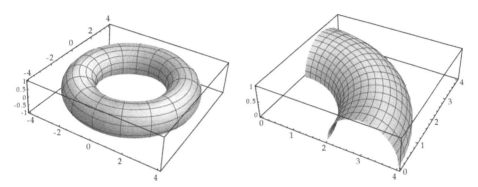

Figure 4.4. The torus $\mathbb{T}^2$ is a differentiable manifold. See the text for the construction of the charts.

4.7 Exercise

Find a set of charts for the torus, $\mathbb{T}^2$, as a differentiable manifold (see figure 4.4).

Solution. The torus of revolution with azimuthal radius c and polar radius a is given by

$$(x, y) \longmapsto (x, y, z) \text{ with } x = (c + a \cos u)\cos v$$
$$y = (c + a \cos u)\sin v$$
$$z = a \sin u$$

Here, u is the polar angle and v is the azimuthal angle. $c > a$ gives us the regular torus (shown above), $c = a$ gives us a torus which is tangent to itself at $(0, 0, 0)$, and $c < a$ gives us a torus which intersects itself. We will consider $c > a$.

To avoid problems at $u = 0$ and $u = 2\pi$, $v = 0$ and $v = 2\pi$, we take two charts

$$\varphi_1 \colon u \in (0, 2\pi) \times v \in (0, 2\pi)$$
$$\varphi_2 \colon u \in (-\pi, \pi) \times v \in (-\pi, \pi)$$

Note that we still have not covered the points $(u = \pi, v = 0)$ and $(u = 0, v = \pi)$. Therefore, we must take two more charts whose domains are open sets containing each of these points. We could deform the domains of φ_1 and φ_2 in such a way that the two uncovered points will drift closer to each other, allowing us to take only one more chart which covers both points. The minimum number of charts needed to cover $\mathbb{T}^2$ is thus 3.

4.8 Definition (*j*th coordinate function)

Let (M, Φ) be an *n*-dimensional differentiable manifold of class C^k, and let $\varphi \in \Phi$ be a coordinate system in $U = \text{dom}(\varphi) \subset M$. Furthermore, let r^j be a function given by

$$r^j \colon \mathbb{R}^n \longrightarrow \mathbb{R}$$
$$(a^1, \ldots, a^n) \longmapsto a^j$$

We define the '*j*th coordinate function' of the coordinate system φ as

$$x^j = r^j \circ \varphi \colon U \longrightarrow \mathbb{R}$$

We have seen several examples of differentiable manifolds. The following lemma will enrich us further by allowing us to construct more sophisticated examples.

4.9 Lemma

Let (M, Φ) be an *m*-dimensional differentiable manifold of class C^k and (N, Ψ) be an *n*-dimensional differentiable manifold of class C^l. $M \times N$, then, has the structure of an $(m + n)$-dimensional differentiable manifold of class C^j, where $j = \min\{k, l\}$.

Proof. Take $p = (x, y) \in M \times N$. By hypothesis, there exist $U \subset M$, $V \subset N$, $\varphi \colon U \longrightarrow \mathbb{R}^m, \psi \colon V \longrightarrow \mathbb{R}^n$ such that $x \in U$, $y \in V$, $\varphi \in C^k$ is a coordinate chart of x and $\psi \in C^l$ is a coordinate chart of y. We define

$$\lambda \colon U \times V \longrightarrow \mathbb{R}^{m+n}$$
$$(x, y) \longmapsto (\varphi(x), \psi(y))$$

Then $\lambda \in C^{\min\{k, l\}}$ and, given $p' \in M \times N$ and λ' defined in an analogous way, $\lambda \circ \lambda'^{-1}$ and $\lambda' \circ \lambda^{-1}$ are functions of class $C^{\min\{k, l\}}$ (inherited from φ, ψ, φ^{-1}, and ψ^{-1}). Furthermore, λ is a homeomorphism, since φ and ψ are homeomorphisms. $\square$

4.10 Definition (differentiable function)

Let (M, Φ) be an *m*-dimensional differentiable manifold of class C^k and $W \subset M$ be an open set. Let also $f \colon W \longrightarrow \mathbb{R}$. Then f is 'of class C^l in W', with $l \leqslant k$, if and only if $\forall p \in W$ there exists a chart φ of a vicinity U of p such that

$$f \circ \varphi^{-1}|_{\varphi(W \cap U)} \in C^l$$

It is common to denote by (U, φ) the chart φ with domain $U \subset M$. We will use this in all that follows.

Chapter 5

Tangent vectors and tangent spaces

To understand what a vector tangent to a manifold is, we must recall that in a Euclidean space a vector naturally defines a differential operator mapping the class of differentiable functions to the real numbers. This operator is the directional derivative with respect to the vector in question.

5.1 Definition (curve on a manifold)

A 'curve of class C^k' on a manifold M is a function

$$\alpha : (a, b) \subset \mathbb{R} \longrightarrow M$$

such that $\forall \, \varphi \in \Phi$ we have $\varphi \circ \alpha : (a, b) \longrightarrow \mathbb{R}^n \in C^k$.

Note that if $f : M \longrightarrow \mathbb{R}$ is of class C^k with $k \geqslant 1$ then

$$f \circ \alpha : I = (a, b) \longrightarrow \mathbb{R}$$
$$t \longmapsto f(\alpha(t))$$

has a well-defined derivative: the rate of change of f along α.

Consider another curve $\beta(s)$ such that α and β pass through a common point $p = \alpha(t_0) = \beta(s_0)$. We say that '$\alpha$ and β have the same velocity at p' if and only if, for all $f : M \longrightarrow \mathbb{R}$,

$$\frac{d}{dt}(f \circ \alpha)\bigg|_{t=t_0} = \frac{d}{ds}(f \circ \beta)\bigg|_{s=s_0} \tag{5.1}$$

5.2 Lemma

Let φ be an arbitrary chart whose domain includes a point p. Suppose $\varphi(p)$ is given by $(x^i(p))_{i=1,\ldots,n}$, where $x^i(p)$ is the ith coordinate of p. Then equation (5.1) is true if and only if, for all $i \in \{1, \ldots, n\}$, we have

$$\left[\frac{d}{dt}(x^i \circ \alpha)\right]_{t=t_0} = \left[\frac{d}{ds}(x^i \circ \beta)\right]_{s=s_0} \tag{5.2}$$

Proof. The fact that (5.1) $\Rightarrow$ (5.2) is trivial.

To prove that (5.2) $\Rightarrow$ (5.1), we write $f \circ \alpha = (f \circ \varphi^{-1}) \circ (\varphi \circ \alpha)$. Now,

$$f \circ \varphi^{-1} : \mathbb{R}^n \longrightarrow \mathbb{R}$$
$$x^i \longmapsto f(\varphi^{-1}(x^i))$$

$$\varphi \circ \alpha \ : I \longrightarrow \mathbb{R}^n$$
$$t \longmapsto \varphi(\alpha(t)) = x^i(\alpha(t))$$

Therefore,

$$\frac{d}{dt}(f \circ \alpha) = \sum_{i=1}^{n} \left[\frac{\partial}{\partial x^i}(f \circ \varphi^{-1})\right] \frac{d}{dt}(\varphi \circ \alpha)$$

$$= \sum_{i=1}^{n} \left[\frac{\partial}{\partial x^i}(f(\varphi^{-1}(x^i)))\right] \frac{d}{dt} x^i(\alpha(t)) \tag{5.3}$$

and similarly for $f \circ \beta$. $\qquad\qquad\square$

5.3 Definition (tangent vector)

At each point $p \in M$, we define

$$\dot{\alpha}_p : C^k(M, \mathbb{R}^n) \longrightarrow \mathbb{R}$$

$$f \longmapsto \left[\frac{d}{dt}(f \circ \alpha)\right]_p = (f \circ \alpha)_p \tag{5.4}$$

for a curve $\alpha(t)$ which passes through p. We then call $\dot{\alpha}_p$ the 'tangent vector to the curve $\alpha(t)$ at p' and we denote it by

$$\dot{\alpha}_p = \left(\frac{d\alpha}{dt}\right)_p$$

Note that vectors are differential operators which act on real-valued functions over M. A vector assigns to a function its directional derivative along a curve to which the vector is tangent. Also note the following:

(i) For a given chart φ with coordinates x^i, the components of $\dot{\alpha}_p$ with respect to φ are

$$\dot{\alpha}_p(x^i) = (x^i \circ \alpha)^\cdot{}_p = \frac{d}{dt} x^i(\alpha(t)) \bigg|_p$$

(ii) Consider two curves $\alpha(t)$ and $\beta(s)$ which pass through a common point p. Then $\dot{\alpha}_p = \dot{\beta}_p \iff (5.1) \overset{\text{lemma 5.2}}{\iff} (5.2) \iff$ the components of $\dot{\alpha}_p$ and $\dot{\beta}_p$ are equal.

(iii) The set of all vectors tangent to M at a point p is called the 'tangent space to M at p' and is denoted by $T_p(M)$.

5.4 Theorem (dimension of the tangent space)

If $\dim(M) = n$, then $T_p(M)$ is an n-dimensional vector space.

Proof. Let X_p, $Y_p \in T_p(M)$, and let $a \in \mathbb{R}$. Then $\exists\, \alpha, \beta : I \longrightarrow M$ such that $\alpha(t_0) = \beta(t_0) = p$ for some t_0 and such that $\dot{\alpha}_p = X_p$ and $\dot{\beta}_p = Y_p$. We define

$$\tilde{\nu} : I \longrightarrow \mathbb{R}^n$$
$$t \longmapsto \varphi \circ \alpha + \varphi \circ \beta$$

$$\nu : I \longrightarrow M$$
$$t \longmapsto (\varphi^{-1} \circ \tilde{\nu})(t) = \varphi^{-1}(\varphi(\alpha(t)) + \varphi(\beta(t)))$$

Then for an arbitrary function $f : M \longrightarrow \mathbb{R}$ we have

$$\dot{\nu}_p(f) = \frac{d}{dt}(f \circ \nu)_p = \frac{d}{dt}[(f \circ \varphi^{-1}) \circ (\varphi \circ \nu)]_p$$

$$\overset{(5.3)}{=} \sum_{i=1}^{n} \left[\frac{\partial}{\partial x^i}(f(\varphi^{-1}(x^i))) \right] \frac{d}{dt} x^i(\nu(t))$$

$$= \sum_{i=1}^{n} \left[\frac{\partial}{\partial x^i}(f(\varphi^{-1}(x^i))) \right] \frac{d}{dt} x^i \varphi^{-1}(\varphi(\alpha(t)) + \varphi(\beta(t)))$$

$$= \sum_{i=1}^{n} \left[\frac{\partial}{\partial x^i}(f(\varphi^{-1}(x^i))) \right] \frac{d}{dt}(\varphi(\alpha(t)) + \varphi(\beta(t)))$$

$$= \sum_{i=1}^{n} \left[\frac{\partial}{\partial x^i}(f(\varphi^{-1}(x^i))) \right] \left[\frac{d}{dt}\varphi(\alpha(t)) + \frac{d}{dt}\varphi(\beta(t)) \right]$$

Therefore,

$$\dot{\nu}_p(f) = \frac{d}{dt}(f \circ \alpha)_p + \frac{d}{dt}(f \circ \beta)_p = X_p f + Y_p f \tag{5.5}$$

that is, there exists $\nu : I \longrightarrow M$ which passes through p such that $\dot{\nu}_p(f) = (X_p + Y_p)f$, and therefore

$$X_p + Y_p \in T_p(M)$$

It is trivial to show that $aX_p \in T_p(M)$. Therefore, $T_p(M)$ has vector-space structure.

We now need to prove that $\dim(T_p(M)) = n$. Let φ be a chart with coordinates $\{x^i\}$. We define n curves α_k as follows:

$$\varphi(\alpha_k(t)) = (x^1(p), \ldots, x^{k-1}(p), x^k(p) + t, x^{k+1}(p), \ldots, x^n(p))$$

In other words, α_k is the image of the coordinate axis in $\mathbb{R}^n$ corresponding to x^k. We take also $p = \alpha_k(0) \ \forall \ k$. Then

$$
\begin{aligned}
(\dot{\alpha}_k f)_p &= \frac{d}{dt}(f \circ \alpha_k)_p = \frac{d}{dt}[(f \circ \varphi^{-1}) \circ (\varphi \circ \alpha_k)]_p \\
&= \sum_{j=1}^{n} \frac{\partial}{\partial x^j}(f \circ \varphi^{-1})\frac{d}{dt}(x^j \circ \alpha_k) \\
&= \sum_{j=1}^{n} \frac{\partial}{\partial x^j}(f \circ \varphi^{-1})\delta_k^j \\
&= \left[\frac{\partial}{\partial x^k}(f \circ \varphi^{-1})\right]_p
\end{aligned}
$$

(5.6)

Equation (5.6) justifies the notation

$$\dot{\alpha}_k(p) = \left(\frac{\partial}{\partial x^k}\right)_p$$

(5.7)

and proves that the $\dot{\alpha}_k$s are linearly independent. Therefore, $\dim(T_p(M)) \geqslant n$. To prove that any vector in $T_p(M)$ is a linear combination of the $\dot{\alpha}_k$ s, we take $X_p \in T_p(M)$ with curve $\alpha(t)$ such that $\alpha(0) = p$. Then for any function $f : M \longrightarrow \mathbb{R}$ we have

$$
\begin{aligned}
X_p(f) &= (f \circ \alpha)'(0) = (f \circ \varphi^{-1} \circ \varphi \circ \alpha)'(0) \\
&= \sum_k \left[\frac{\partial}{\partial x^k}f \circ \varphi^{-1}\right](x^k \circ \alpha)'(0) \\
&\overset{(5.6)}{=} \sum_k \left[\left(\frac{\partial}{\partial x^k}\right)_p f\right] X_p(x^k)
\end{aligned}
$$

(5.8)

so that

$$X_p = \sum_k (X_p(x^k))\left(\frac{\partial}{\partial x^k}\right)_p$$

(5.9)

as expected. In other words, $\dim(T_p(M)) = n$. $\qquad\square$

The above proof has the advantage of being constructive: from (5.9), we see that $\{X_p(x^i)\}$ are the components of X_p with respect to the basis $\{\dot{\alpha}_k\}$. We must distinguish between the differential operator $\frac{\partial}{\partial x^k}$ on $\mathbb{R}^n$ and the vector $(\frac{\partial}{\partial x^k})_p \in T_p(M)$. However, note that equation (5.6) relates one to the other.

The union of tangent spaces $T_p(M)$ at all points of a manifold M is called the 'tangent bundle of M' and is denoted by

$$T(M) = \bigcup_{p \in M} T_p(M)$$

Observe that $T(M)$ is *not* a vector space: we cannot add vectors which are tangent to M at different points. To do so, we would need a way to transport a vector throughout M. There are several ways to do this, and we shall introduce them in subsequent chapters while analysing their geometric meaning.

5.5 Lemma (Leibnitz's rule)

Let $f, g : M \longrightarrow \mathbb{R}$, and we define

$$f\,g : M \longrightarrow \mathbb{R}$$
$$p \longmapsto f(p)\,g(p)$$

Let $X_p \in T_p(M)$. Then

$$X_p(f\,g) = X_p(f)\,g(p) + f(p)\,X_p(g)$$

Proof. Let α be a curve on M such that X_p is tangent to α at $t = 0$. Then

$$X_p(f\,g) = [(f\,g) \circ \alpha]^{\cdot}_{t=0} = \frac{d}{dt}[(f\,g) \circ \varphi^{-1} \circ \varphi \circ \alpha]_{t=0}$$

$$= \sum_k \left[\frac{\partial}{\partial x^k}((f\,g) \circ \varphi^{-1})\right]_p \left[\frac{d}{dt}(x^k \circ \alpha)\right]_{t=0}$$

$$= \sum_k \left[\frac{\partial}{\partial x^k}(f\,g)\right]_p \left[\frac{d}{dt}(x^k \circ \alpha)\right]_{t=0}$$

$$= \sum_k \left[\left(\frac{\partial}{\partial x^k}f\right)_p g(p) + f(p)\left(\frac{\partial}{\partial x^k}g\right)_p\right]\left[\frac{d}{dt}(x^k \circ \alpha)\right]_{t=0}$$

$$= \sum_k \left[\frac{\partial}{\partial x^k}(f \circ \varphi^{-1})\right]_p \left[\frac{d}{dt}(x^k \circ \alpha)\right]_{t=0} g(p)$$

$$+ \sum_k \left[\frac{\partial}{\partial x^k}(g \circ \varphi^{-1})\right]_p \left[\frac{d}{dt}(x^k \circ \alpha)\right]_{t=0} f(p)$$

$$= X_p(f)\,g(p) + f(p)\,X_p(g)$$

$\square$

5.6 Exercise (relative curvature)

Consider two curves $\alpha(t)$ and $\beta(t)$ of class C^k which are tangent to each other at a point p given by $t = 0$, with the same speed; in other words, $\dot{\alpha}(0) = \dot{\beta}(0)$.

Show that the functional

$$a : C^k(U, \mathbb{R}) \longrightarrow \mathbb{R}$$
$$f \longmapsto (f \circ \beta - f \circ \alpha)^{\cdot\cdot}(0)$$

(where U is a vicinity of p) is a tangent vector, that is, that $a \in T_p(M)$. (This vector is called the 'relative curvature' of β with respect to α.)

Solution.

$$(f \circ \alpha)^{\cdot\cdot} = \frac{d}{dt}(f \circ \alpha)^{\cdot} = \sum_i \frac{d}{dt}\left(\frac{\partial f}{\partial x^i}\frac{d}{dt}x^i(\alpha(t))\right)$$

$$= \sum_{i,j}\left[\frac{\partial f}{\partial x^i}\frac{d^2}{dt^2}x^i(\alpha(t)) + \frac{\partial^2 f}{\partial x^j \partial x^i}\frac{d}{dt}x^j(\alpha(t))\frac{d}{dt}x^i(\alpha(t))\right]$$

Thus $(f \circ \alpha)^{\cdot\cdot}$ is not a vector, since it involves second derivatives of f. But at p we have

$$\frac{dx^i}{dt}\alpha(t) = \frac{dx^i}{dt}\beta(t)$$

Therefore,

$$(f \circ \beta - f \circ \alpha)^{\cdot\cdot}{}_p = \sum_i \frac{\partial f}{\partial x^i}\left[\frac{d^2}{dt^2}x^i(\beta(t)) - \frac{d^2}{dt^2}x^i(\alpha(t))\right]_p$$

$$+ \sum_{i,j}\frac{\partial^2 f}{\partial x^i \partial x^j}\left[\frac{d}{dt}x^i(\beta(t))\frac{d}{dt}x^j(\beta(t)) - \frac{d}{dt}x^i(\alpha(t))\frac{d}{dt}x^j(\alpha(t))\right]_p$$

$$= \sum_i \frac{\partial f}{\partial x^i}\frac{d^2}{dt^2}(x^i(\beta(t)) - x^i(\alpha(t)))_p$$

In other words, $f \longmapsto (f \circ \beta - f \circ \alpha)^{\cdot\cdot}{}_p$ defines a tangent vector at p whose components are

$$\frac{d^2}{dt^2}(x^i(\beta) - x^i(\alpha))_p$$

Defining a vector field on a manifold M (i.e. assigning a tangent vector at each point of M) is equivalent to describing a dynamic system on M, as we will see in the next example.

5.7 Example (dynamic system)

Let M be a two-dimensional differentiable manifold, and let (x, y) be the coordinates of a chart φ in M. Furthermore, consider $v \in T(M)$ given by

$$v = y\frac{\partial}{\partial x} - (y + x)\frac{\partial}{\partial y}$$

so that

$$v(x) = y$$
$$v(y) = -(y + x)$$

(5.10)

Let $\dot{\alpha} : I \longrightarrow M$ be an integral curve of v, that is,

$$v(x) = (x \circ \alpha)^{\cdot} = \frac{d}{dt} x(\alpha(t))$$

$$v(y) = (y \circ \alpha)^{\cdot} = \frac{d}{dt} y(\alpha(t))$$

(5.11)

Combining (5.10) and (5.11), we have

$$\frac{d^2}{dt^2} x(\alpha(t)) = \frac{d}{dt} y(\alpha(t)) = -[y(\alpha) + x(\alpha)] = -\frac{d}{dt} x(\alpha) - x(\alpha)$$

In other words,

$$\frac{d^2}{dt^2} x(\alpha) + \frac{d}{dt} x(\alpha(t)) + x(\alpha) = 0$$

which represents a damped harmonic oscillator with the following equation of motion:

$$x(t) = e^{-\frac{t}{2}} \left[A_1 \cos\left(\frac{\sqrt{3}}{2} t\right) + A_2 \sin\left(\frac{\sqrt{3}}{2} t\right) \right]$$

Its behaviour is plotted in figure 5.1.

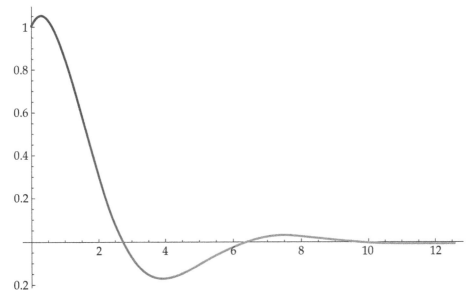

Figure 5.1. A damped harmonic oscillator defined by the vector field $v = y \frac{\partial}{\partial x} - (y + x)\frac{\partial}{\partial y}$.

Chapter 6

Tensor algebra

We now know that $T_p(M)$ is a vector space. We can therefore use on it the machinery of linear algebra. In this chapter we will introduce the concept of 'tensor', which is but a multilinear operator over a vector space. As such, it is independent of coordinate systems (only its components depend on the chosen system). The physical quantities which are actually real are, then, tensors (with scalars, vectors and 1-forms as particular cases).

6.1 Definition (1-forms)

The 'dual space' $T_p^*(M)$ of $T_p(M)$ is the space of linear functions

$$\lambda \colon T_p(M) \longrightarrow \mathbb{R}$$

$T_p^*(M)$ is also an n-dimensional vector space (see lemma 6.3). The elements of $T_p^*(M)$ are called '1-forms' or sometimes just '(differential) forms'. Some authors still call them 'covectors'.
In what follows, we will use the Einstein notation (see section 1.6).

6.2 Example (differential of a function)

Let $f : M \longrightarrow \mathbb{R}$ be an arbitrary function. Then $X_p(f)$ is a scalar $\forall X_p \in T_p(M)$. In other words, f defines a map

$$\begin{aligned} df : T_p(M) &\longrightarrow \mathbb{R} \\ X_p &\longmapsto X_p(f) \end{aligned} \tag{6.1}$$

Since X_p is linear, so is df, which is to say that $df \in T_p^*(M)$. df is called the 'differential of f' or the 'gradient of f'.

6.3 Lemma (dual basis of T_p^*)

Let φ be a chart with coordinates x^i. Then $\{dx^i\}$ is a basis of T_p^* with the property that

$$dx^i\left(\frac{\partial}{\partial x^j}\right)_p = \delta_j^i \tag{6.2}$$

Proof.

$$dx^i\left(\frac{\partial}{\partial x^j}\right)_p \overset{(6.1)}{=} \left(\frac{\partial}{\partial x^j}\right)_p x^i \overset{(5.6)}{=} \left(\frac{\partial}{\partial x^j}\right)(x^i \circ \varphi^{-1})_{\varphi(p)} = \delta_j^i$$

To prove that $\{dx^i\}$ is linearly independent, take an arbitrary linear combination $\omega_i\, dx^i = 0$. Then

$$\omega_i\left[dx^i\left(\frac{\partial}{\partial x^j}\right)_p\right] = 0$$

By equation (6.2), the left side of the preceding equation is simply ω_j. Therefore, $\omega_j = 0 \;\forall\, j$, so $\{dx^i\}$ is linearly independent.

To prove that any 1-form may be written as a linear combination of the dx^i, take an arbitrary $\lambda \in T_p^*(M)$ and define

$$\mu = \lambda - \lambda\left(\frac{\partial}{\partial x^i}\right)_p dx^i \tag{6.3}$$

Then $\forall\, j \in \{1, \ldots, n\}$ we have

$$\mu\left(\frac{\partial}{\partial x^j}\right)_p = \lambda\left(\frac{\partial}{\partial x^j}\right)_p - \lambda\left(\frac{\partial}{\partial x^i}\right)_p dx^i\left(\frac{\partial}{\partial x^j}\right)_p \overset{(6.2)}{=} 0$$

or, in other words,

$$\lambda = \lambda\left(\frac{\partial}{\partial x^j}\right)_p dx^i \tag{6.4}$$

$\square$

6.4 Notation

(i) We shall use the following notation for a 1-form acting on a vector:

$$\langle \lambda, X_p \rangle \equiv \lambda(X_p)$$

(ii) Equation (6.4) shows that any 1-form may be written as

$$\lambda = \left\langle \lambda, \left(\frac{\partial}{\partial x^i}\right)_p \right\rangle dx^i$$

In particular, for $\lambda = df$ we have

$$\left\langle df, \left(\frac{\partial}{\partial x^i}\right)_p \right\rangle = \left(\frac{\partial}{\partial x^i}\right)_p f \overset{(5.6)}{=} \left(f_{,i}\right)_p$$

where $f_{,i} = \frac{\partial}{\partial x^i}(f \circ \varphi^{-1})$ or, put differently,

$$df = \left(f_{,i}\right)_p dx^i \tag{6.5}$$

and hence the name 'gradient of f'.

6.5 Corollary (change of basis)

(i) Given a coordinate transformation (that is, a switch from one chart to another) $x^i \longmapsto y^i(x^j)$, equation (6.5) implies that

$$dy^i = A^i{}_j dx^j \tag{6.6}$$

where

$$A^i{}_j = \left(\frac{\partial y^i}{\partial x^j}\right)_p$$

(ii) If $(\frac{\partial}{\partial y^i})_p$ is the new basis of $T_p(M)$ (it is the dual basis of $\{dy^i\}$), then from equation (6.2) we obtain

$$\left(\frac{\partial}{\partial y^j}\right)_p = (A^{-1})^i{}_j \left(\frac{\partial}{\partial x^i}\right)_p \tag{6.7}$$

where

$$(A^{-1})^i{}_j = \left(\frac{\partial x^i}{\partial y^j}\right)_p$$

Equations (6.6) and (6.7) give us the formulae for a change of basis. Notice the position of the indices, and remember that an upper index in a denominator is equivalent to a lower index in a numerator.

6.6 Corollary (generalisation of corollary 6.5)

Let $\{e_a\}$ be a basis of $T_p(M)$ and $\{\omega^a\}$ be its dual basis in $T_p^*(M)$, that is,

$$\langle \omega^a, e_b \rangle = \delta_b^a \qquad (a, b = 1, \dots, n) \tag{6.8}$$

Then the formulae for a change of basis $\omega^a \longmapsto \hat{\omega}^a$, $e_a \longmapsto \hat{e}_a$ are

$$\hat{\omega}^a = A^a{}_b\,\omega^b, \qquad \hat{e}_a = (A^{-1})^b{}_a\,e_b \tag{6.9}$$

where $A^a{}_b$ is a nonsingular matrix.

6.7 Transformation rules

Let λ be a 1-form with components λ_a with respect to the basis $\{\omega^a\}$. Then

$$\lambda = \lambda_a\,\omega^a = \hat{\lambda}_a\,\hat{\omega}^a \stackrel{(6.9)}{=} \hat{\lambda}_a\,A^a{}_b\,\omega^b$$

Since $\{\omega^a\}$ is linearly independent,

$$\lambda_b = A^a{}_b\,\hat{\lambda}_a, \qquad \hat{\lambda}_a = (A^{-1})^b{}_a\,\lambda_b \tag{6.10}$$

Similarly, if X_p is a vector ($X_p = X^a\,e_a = \hat{X}^a\,\hat{e}_a$), then, by equation (6.9),

$$X^b = (A^{-1})^b{}_a\,\hat{X}^a, \qquad \hat{X}^a = A^a{}_b\,X^b \tag{6.11}$$

Note the differences between equations (6.10) and (6.11). Quantities which are transformed as those in equation (6.10) are called 'covariant', while quantities which are transformed as those in equation (6.11) are called 'contravariant'.

6.8 Definition (tensor)

Let $S\colon T_p(M) \times T_p(M) \times T_p^*(M) \longrightarrow \mathbb{R}$ be a map that is linear in each of its arguments. S is then called a 'tensor of rank $\binom{1}{2}$ on M' or simply a '$\binom{1}{2}$ tensor'.

6.9 Lemma

Let S be a tensor of rank $\binom{1}{2}$ on M, and let $\{e_a\}$ and $\{\omega^a\}$ be dual bases on M. By definition, we have

$$S_{ab}{}^c \stackrel{\text{def}}{=} S(e_a, e_b, \omega^c) \in \mathbb{R}$$

Then $\forall\, X,\, Y \in T_p(M),\ \forall\, \lambda \in T_p^*(M)$, we have

$$S(X,\, Y,\, \lambda) = S_{ab}{}^c\, X^a\, Y^b\, \lambda_c$$

(that is, S is determined by its components with respect to a basis).

Proof. Take arbitrary $X,\, Y \in T_p(M),\ \lambda \in T_p^*(M)$. We write $X = X^a\,e_a$, $Y = Y^a\,e_a$, $\lambda = \lambda_a\,\omega^a$. Then

$$\begin{aligned}
S(X,\, Y,\, \lambda) &= S(X^a\,e_a,\, Y^b\,e_b,\, \lambda_c\,\omega^c)\\
&\stackrel{\text{linear}}{=} X^a\, Y^b\, \lambda_c\, S(e_a,\, e_b,\, \omega^c)\\
&= X^a\, Y^b\, \lambda_c\, S_{ab}{}^c
\end{aligned}$$

$\square$

The generalisation of definition (6.8) and lemma (6.9) to tensors of rank $\binom{r}{s}$, with arbitrary $r, s \in \mathbb{N} \cup \{0\}$, is straightforward. It is also clear that the sum or product of two $\binom{r}{s}$ tensors is another $\binom{r}{s}$ tensor. Therefore, the set of $\binom{r}{s}$ tensors is a vector space. However, the sum of, say, a $\binom{0}{1}$ tensor (a 1-form) and a $\binom{4}{7}$ tensor is undefined.

6.10 Definition (tensor product)

(i) Let S be a $\binom{1}{2}$ tensor and T be a $\binom{1}{1}$ tensor. Then

$$U: T_p \times T_p \times T_p^* \times T_p \times T_p^* \longrightarrow \mathbb{R}$$
$$(X, Y, \lambda, Z, \mu) \longmapsto S(X, Y, \lambda)T(Z, \mu)$$

is called the 'tensor product' of S and T and is denoted by $U = S \otimes T$. Note that U is a $\binom{2}{3}$ tensor whose components are $U_{ab}{}^c{}_d{}^e = S_{ab}{}^c T_d{}^e$. The product of an $\binom{r}{s}$ tensor with a $\binom{p}{q}$ tensor is an $\binom{r+p}{s+q}$ tensor.

(ii) Let U be as in (i), and let $V_{ab}{}^e = U_{ab}{}^c{}_c{}^e$. The (unique) tensor V of rank $\binom{1}{2}$ and components $V_{ab}{}^e$ is then called the 'contraction of U on the third and fourth indices'. The contraction on any other pair of indices is defined similarly provided one of the indices is contravariant (on top) and the other is covariant (on the bottom).

(iii) Let S be as in (i). Then

$$S_{(ab)}{}^c = \frac{1}{2!}(S_{ab}{}^c + S_{ba}{}^c) \qquad \text{and} \qquad S_{[ab]}{}^c = \frac{1}{2!}(S_{ab}{}^c - S_{ba}{}^c)$$

are new tensors called, respectively, the 'symmetric part of S' and the 'antisymmetric part of S' on the first two indices. The generalisation to tensors of any rank and any set of indices is straightforward. For example:

$$T_{[abc]} = \frac{1}{3!}(T_{abc} + T_{bca} + T_{cab} - T_{acb} - T_{bac} - T_{cba})$$

6.11 Definition (symmetrisation)

A tensor which is symmetric in all its indices is called a 'symmetric tensor'. A tensor which is antisymmetric in all its indices is called an 'antisymmetric tensor'.

6.12 Exercise

(i) What is the transformation law, corresponding to equations (6.10) and (6.11), for a tensor?

Solution.

$$S^{\beta_1\beta_2\cdots}_{\alpha_1\alpha_2\cdots} = A^{a_1}_{\ \alpha_1} A^{a_2}_{\ \alpha_2} \cdots (A^{-1})^{\beta_1}_{\ b_1} (A^{-1})^{\beta_2}_{\ b_2} \cdots \hat{S}^{b_1b_2\cdots}_{a_1a_2\cdots}$$

(ii) Suppose the components of two tensors are equal in a given basis. Prove that they are then equal in all bases.

Solution. Suppose $S^{b_1b_2\cdots}_{a_1a_2\cdots} = T^{b_1b_2\cdots}_{a_1a_2\cdots}$ in a basis $\{e_\alpha, \omega^\alpha\}$. We perform a basis transformation:

$$\hat{S}^{\beta_1\beta_2\cdots}_{\alpha_1\alpha_2\cdots} = A^{\beta_1}_{\ b_1} A^{\beta_2}_{\ b_2} \cdots (A^{-1})^{a_1}_{\ \alpha_1} (A^{-1})^{a_2}_{\ \alpha_2} \cdots S^{b_1b_2\cdots}_{a_1a_2\cdots}$$

$$= A^{\beta_1}_{\ b_1} A^{\beta_2}_{\ b_2} \cdots (A^{-1})^{a_1}_{\ \alpha_1} (A^{-1})^{a_2}_{\ \alpha_2} \cdots T^{b_1b_2\cdots}_{a_1a_2\cdots}$$

$$= \hat{T}^{\beta_1\beta_2\cdots}_{\alpha_1\alpha_2\cdots}$$

6.13 Exercise

A particular $\begin{pmatrix} 0 \\ 4 \end{pmatrix}$ tensor has the following symmetry properties:

$$R_{ab(cd)} = 0, \quad R_{(ab)cd} = 0, \quad R_{a[bcd]} = 0$$

Show that $R_{abcd} = R_{cdab}$.

Solution.
    (i) $R_{a[bcd]} = 0 \implies R_{abcd} + R_{acdb} + R_{adbc} - R_{abdc} - R_{acbd} - R_{adcb} = 0$.
    (ii) $R_{(ab)cd} = 0 \implies R_{abcd} + R_{bacd} = 0$.
    (iii) $R_{ab(cd)} = 0 \implies R_{abcd} + R_{abdc} = 0$.
    (iv) From (i) and (iii), we obtain $R_{abcd} + R_{acdb} + R_{adbc} = 0$.

Therefore,

$$\begin{aligned}
R_{abcd} &= -R_{acdb} - R_{adbc} \stackrel{(ii)}{=} R_{cadb} + R_{dabc} \\
&\stackrel{(iv)}{=} -R_{cbad} - R_{cdba} - R_{dbca} - R_{dcab} \\
&\stackrel{(ii),(iii)}{=} R_{bcad} + R_{bdca} + R_{cdab} + R_{cdab} \\
&\stackrel{(iv)}{=} -R_{badc} + 2R_{cdab} \\
&\stackrel{(ii),(iii)}{=} -R_{abcd} + 2R_{cdab}
\end{aligned}$$

The desired result follows immediately from this.

6.14 Exercise

Let A be a $\begin{pmatrix} 0 \\ 2 \end{pmatrix}$ tensor whose components with respect to a locally Cartesian basis at a point p are

$$A_{ij} = \begin{bmatrix} 1 & 2 & 0 \\ 0 & 1 & 2 \\ 2 & 0 & 1 \end{bmatrix}$$

(i) Separate A_{ij} into a symmetric part T_{ij} and an antisymmetric part W_{ij}.

Solution.

$$T_{ij} = \frac{1}{2}\left(A_{ij} + A_{ji}\right) = \begin{bmatrix} 1 & 1 & 1 \\ 1 & 1 & 1 \\ 1 & 1 & 1 \end{bmatrix}$$

$$W_{ij} = \frac{1}{2}\left(A_{ij} - A_{ji}\right) = \begin{bmatrix} 0 & 1 & -1 \\ -1 & 0 & 1 \\ 1 & -1 & 0 \end{bmatrix}$$

(ii) Diagonalise T_{ij} and obtain the direction cosines for a principal vector basis at p. Describe the quadratic surface $T_{ij}\, x^i\, x^j = $ const.

Solution. Note that T_{ij} is singular. To obtain its eigenvalues, we write

$$0 = \det(T_{ij} - \lambda\, I_{ij}) = \begin{vmatrix} 1-\lambda & 1 & 1 \\ 1 & 1-\lambda & 1 \\ 1 & 1 & 1-\lambda \end{vmatrix} = (1-\lambda)^3 + 2 - 3(1-\lambda)$$

Therefore, $0 = \lambda^3 - 3\lambda^2 = \lambda^2(\lambda - 3)$, that is, T_{ij}'s eigenvalues are

$$\lambda_1 = 3, \qquad \lambda_2 = 0, \qquad \lambda_3 = 0$$

The diagonal form of T_{ij} is then

$$\begin{bmatrix} 3 & 0 & 0 \\ 0 & 0 & 0 \\ 0 & 0 & 0 \end{bmatrix}$$

To obtain T_{ij}'s eigenvectors, we write

$$(a_1 + a_2 + a_3,\ a_1 + a_2 + a_3,\ a_1 + a_2 + a_3) = T_{ij}\, a^j = 3\, a_i = (3\, a_1,\ 3\, a_2,\ 3\, a_3)$$

Therefore, $a_1 = a_2 = a_3$. Normalising, we obtain

$$a_i = \sqrt{\frac{1}{3}} \quad (i = 1, 2, 3)$$

that is,

$$\vec{a} = \left(\sqrt{\frac{1}{3}}, \sqrt{\frac{1}{3}}, \sqrt{\frac{1}{3}}\right)$$

For the other eigenvectors, we have

$$0 = T_{ij}\, b^j = (b_1 + b_2 + b_3,\ b_1 + b_2 + b_3,\ b_1 + b_2 + b_3)$$

that is, $b_1 = -b_2 - b_3$; any nontrivial vector satisfying this condition is an eigenvector of T_{ij}. Normalising, we have $b_1^2 + b_2^2 + b_3^2 = 1$ and therefore $2\,b_2^2 + 2\,b_3^2 + 2\,b_2\,b_3 = 1$. Choosing $b_3 = 0$, we have

$$\vec{b} = \left(-\sqrt{\frac{1}{2}},\ \sqrt{\frac{1}{2}},\ 0 \right)$$

The third eigenvector may be obtained as follows:

$$\vec{c} = \vec{a} \times \vec{b} = \left(-\sqrt{\frac{1}{6}},\ -\sqrt{\frac{1}{6}},\ 2\sqrt{\frac{1}{6}} \right)$$

Taking $\{\vec{a},\ \vec{b},\ \vec{c}\}$ as a principal basis for T_{ij}, the direction cosines are

$$\alpha = \vec{a} \cdot \hat{i} = \sqrt{\frac{1}{3}}$$

$$\beta = \vec{b} \cdot \hat{j} = \sqrt{\frac{1}{2}}$$

$$\gamma = \vec{c} \cdot \hat{k} = \sqrt{\frac{1}{6}}$$

We calculate the surface $T_{ij}\, x^i x^j = \text{const.}$ in the principal basis for simplicity (i.e. using T_{ij}'s diagonal form):

$$3\, x_1'^2 = \text{const}$$

that is,

$$x_1' = \pm\, \text{const}$$

which represents a pair of planes parallel to the x_2'- and x_3'-axes.

IOP Publishing

Differential Topology and Geometry with Applications to Physics

Eduardo Nahmad-Achar

Chapter 7

Tensor fields and commutators

So far, we have studied the algebra of tensors defined at a single point $p \in M$. In chapter 5 we introduced the idea of the union of tangent spaces $T_p(M)$ at different points of M,

$$T(M) = \bigcup_{p \in M} T_p(M)$$

$T(M)$ is known as the 'tangent bundle of M' and will allow us to extend the notion of a tensor at a point to a tensor field defined on the entire manifold, which is necessary in order to do tensor analysis.

7.1 Definition (vector field)

A function

$$X : M \longrightarrow T(M)$$
$$p \longmapsto X_p \in T_p(M)$$

is called a 'vector field on M'. X specifies a vector at each point of M. Given a coordinate system $\{x^i\}$ and an associated basis $(\frac{\partial}{\partial x^i})_p$ on each $T_p(M)$, X has components $X^i = X(x^i)$:

$$X_p = X\big(x^i(p)\big)\left(\frac{\partial}{\partial x^i}\right)_p \tag{7.1}$$

If the $X(x^i)$ are of class C^k for all the coordinate charts, we say that X is of class C^k.

This definition is readily extended to 1-forms and, in general, to tensors of rank $\binom{r}{s}$.

Let X, Y be vector fields on M, and let $f : M \longrightarrow \mathbb{R}$. For all $p \in M$, we have $Y_p(f) \in \mathbb{R}$. Then

$$Yf : M \longrightarrow \mathbb{R}$$
$$p \longmapsto Y_p(f)$$

is a well-defined function. Therefore, the expression $X_p(Yf)$ makes sense. In coordinates,

$$X_p(Yf) = X^i \left(\frac{\partial}{\partial x^i}\right)_p Yf = X^i \left(\frac{\partial}{\partial x^i}\right)_p \left(Y^j \frac{\partial f}{\partial x^j}\right)$$

$$= X^i Y^j \left(\frac{\partial^2 f}{\partial x^i \partial x^j}\right)_p + X^i \left(\frac{\partial Y^j}{\partial x^i}\right)_p \left(\frac{\partial f}{\partial x^j}\right)_p$$

(7.2)

Therefore, there does *not* exist $Z \in T(M)$ such that $Z_p f = X_p(Yf)$, since equation (7.2) has second derivatives of f. But

$$X_p(Yf) - Y_p(Xf) \overset{(7.2)}{=} X^i Y^j_{,i} \left(\frac{\partial f}{\partial x^j}\right)_p - Y^i X^j_{,i} \left(\frac{\partial f}{\partial x^j}\right)_p$$

(7.3)

is a vector. (The notation $_{,i}$ for partial derivatives was introduced in section 6.4.)

7.2 Definition (commutator)

Let X, Y be vector fields on M, and let $f : M \longrightarrow (R)$. Then

$$[X, Y]f \overset{\text{def}}{=} X(Yf) - Y(Xf)$$

(7.4)

is called the 'commutator of X and Y'. Its components are, as per equation (7.3),

$$[X, Y]^j = X^i Y^j_{,i} - Y^i X^j_{,i}$$

(7.5)

and $[X, Y]$ is a vector field on M. We say that the vector fields X and Y 'commute' if and only if $[X, Y] = 0$.

7.3 Exercise

Using definition 7.2, prove the following:
   (i) $[X, Y] = -[Y, X]$.
   (ii) $[X + Z, Y] = [X, Y] + [Z, Y]$.
   (iii) $[[X, Y], Z] + [[Z, X], Y] + [[Y, Z], X] = 0$.

Identity (iii) is called the 'Jacobi identity'.

7.4 Exercise (structure functions)

Let $\{e_a\}$ be a basis of $T_p(M)$. Since $[e_a, e_b] \in T_p(M)$ we may write $[e_a, e_b] = \gamma^c_{ab} e_c$. Are γ^c_{ab} the components of a tensor?

Solution. Let $\{\omega^a\}$ be the dual basis of $\{e_a\}$, and let $\{\hat{e}_a\}$ and $\{\hat{\omega}^a\}$ be another pair of dual bases, that is, $\hat{\omega}^a = A^a_{\ b} \omega^b$, and $\hat{e}_a = (A^{-1})^b_{\ a} e_b$. Let $f : M \longrightarrow \mathbb{R}$ be an arbitrary function. Then

$$\hat{e}_a(\hat{e}_b f) = (A^{-1})^p{}_a e_p\left((A^{-1})^q{}_b e_q f\right)$$

$$= (A^{-1})^p{}_a (A^{-1})^q{}_b e_p \, e_q f + (A^{-1})^p{}_a e_p\left((A^{-1})^q{}_b\right) e_q f$$

$$= (A^{-1})^p{}_a (A^{-1})^q{}_b e_p \, e_q f + \hat{e}_a(A^{-1})^q{}_b e_q f$$

Therefore,

$$[\hat{e}_a, \hat{e}_b] f = (A^{-1})^p{}_a (A^{-1})^q{}_b [e_p, e_q] f + \hat{e}_a\left[(A^{-1})^q{}_b\right] e_q f - \hat{e}_b\left[(A^{-1})^p{}_a\right] e_p f$$

On the other hand,

$$[\hat{e}_a, \hat{e}_b] f = \hat{\gamma}^c{}_{ab} \, \hat{e}_c \, f = \hat{\gamma}^c{}_{ab} (A^{-1})^s{}_c e_s \, f$$

Therefore,

$$\hat{\gamma}^c{}_{ab} (A^{-1})^s{}_c = (A^{-1})^p{}_a (A^{-1})^q{}_b \gamma^s{}_{pq} + \hat{e}_a (A^{-1})^s{}_b - \hat{e}_b (A^{-1})^s{}_a$$

which is to say that

$$\hat{\gamma}^c{}_{ab} = A^c{}_s (A^{-1})^p{}_a (A^{-1})^q{}_b \gamma^s{}_{pq} + A^c{}_s \, [\hat{e}_a (A^{-1})^s{}_b - \hat{e}_b (A^{-1})^s{}_a]$$

which is not the usual form in which a tensor transforms, that is, the $\gamma^c{}_{ab}$ are *not* the components of a tensor. They are called the 'structure functions'.

Given a coordinate system x^i, it is natural to choose $\left\{\frac{\partial}{\partial x^i}\right\}$ as the basis of the tangent space. However, this is not necessary; any set of n linearly independent vector fields is, in principle, equally useful. Do all vector bases come from a coordinate system? If not, how do we tell those which do from those which do not?

7.5 Definition (space of vector fields, and coordinate basis)

(i) $\mathfrak{X}(M) \overset{\text{def}}{=} \{X \mid X \text{ is a vector field on } M\}$ is the 'space of vector fields on M'. (Observe that, unlike $T(M)$, $\mathfrak{X}(M)$ *is* a vector space with point-by-point scalar multiplication and vector-addition operations:

$$M \longrightarrow T(M)$$
$$p \longmapsto X_p + Y_p \in T_p(M)$$

etc.)

(ii) Let $\{V_1, \ldots, V_n\}$ be a basis of $\mathfrak{X}(M)$. $\{V_1, \ldots, V_n\}$ is called a 'coordinate basis' if and only if, for all $i \in \{1, \ldots, n\}$, V_i can be written as a derivative with respect to a coordinate.

7.6 Theorem (coordinate bases)

A basis $\{V_1, \ldots, V_n\}$ of $\mathfrak{X}(M)$ is a coordinate basis if and only if

$$[V_i, V_j] = 0$$

for all $i, j \in \{1, \ldots, n\}$.

Proof. Let V, $W \in \{V_1, \ldots, V_n\}$, and let $\{x^i\}$ be a coordinate system. Since $\left\{\frac{\partial}{\partial x^i}\right\}$ is also a basis of $\mathfrak{X}(M)$, we have

$$
\begin{aligned}
[V, W] &= V^i \frac{\partial}{\partial x^i} W^j \frac{\partial}{\partial x^j} - W^j \frac{\partial}{\partial x^j} V^i \frac{\partial}{\partial x^i} \\
&= V^i W^j \left(\frac{\partial}{\partial x^i} \frac{\partial}{\partial x^j} - \frac{\partial}{\partial x^j} \frac{\partial}{\partial x^i} \right) + V^i \frac{\partial W^j}{\partial x^i} \frac{\partial}{\partial x^j} - W^j \frac{\partial V^i}{\partial x^j} \frac{\partial}{\partial x^i} \quad (7.6) \\
&= \left(V^i \frac{\partial W^j}{\partial x^i} - W^i \frac{\partial V^j}{\partial x^i} \right) \frac{\partial}{\partial x^j}
\end{aligned}
$$

It is easy to see that the coefficient of $\frac{\partial}{\partial x^i}$ in equation (7.6) is zero when V and W belong to a coordinate basis. Suppose there exist coordinates $\{u^i\}_{i \in \{1, \ldots, n\}}$ such that

$$
V = \frac{\partial}{\partial u^1}, \qquad W = \frac{\partial}{\partial u^2}
$$

Then in this coordinate system we have

$$
V^i_{\{u\}} = \delta^i_1, \qquad W^i_{\{u\}} = \delta^i_2
$$

and therefore

$$
V^i = (A^i_j) V^j_{\{u\}} = \frac{\partial x^i}{\partial u^j} \delta^j_1 = \frac{\partial x^i}{\partial u^1}
$$

$$
W^j = (A^j_k) W^k_{\{u\}} = \frac{\partial x^j}{\partial u^k} \delta^k_2 = \frac{\partial x^j}{\partial u^2}
$$

Thus we have

$$
V^i \frac{\partial W^j}{\partial x^i} = \frac{\partial x^i}{\partial u^1} \frac{\partial}{\partial x^i} \frac{\partial x^j}{\partial u^2} = \frac{\partial}{\partial u^1} \frac{\partial x^j}{\partial u^2} = \frac{\partial^2 x^j}{\partial u^1 \partial u^2}
$$

$$
W^i \frac{\partial V^j}{\partial x^i} = \frac{\partial x^i}{\partial u^2} \frac{\partial}{\partial x^i} \frac{\partial x^j}{\partial u^1} = \frac{\partial}{\partial u^2} \frac{\partial x^j}{\partial u^1} = \frac{\partial^2 x^j}{\partial u^2 \partial u^1} = \frac{\partial^2 x^j}{\partial u^1 \partial u^2}
$$

from which

$$
[V, W] = 0
$$

$\square$

What is the geometric difference between vector fields which come from a coordinate basis and vector fields which come from a non-coordinate basis?

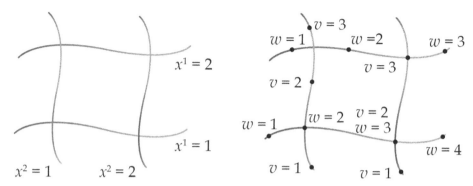

Figure 7.1. Difference between a coordinate basis (left) and a non-coordinate basis (right). In a non-coordinate basis one variable is not constant along the lines of the other variables.

In a coordinate lattice over a two-dimensional manifold (see figure 7.1, left) x^1 is by definition constant along the lines of x^2 (that is, the integral lines of $\frac{\partial}{\partial x^2}$). To put it differently, $\frac{\partial}{\partial x^2}$ is a derivative along a curve where x^1 is constant, and vice versa. Therefore, $\frac{\partial}{\partial x^1}$ and $\frac{\partial}{\partial x^2}$ commute.

Given two arbitrary vector fields $V = \frac{\partial}{\partial v}$ and $W = \frac{\partial}{\partial w}$ (see figure 7.1, right), an integral curve of W is not necessarily a curve where v is constant, that is, the derivative $\frac{\partial}{\partial w}$ does not maintain v invariant (or vice versa), so $\frac{\partial}{\partial v}$ and $\frac{\partial}{\partial w}$ do not commute. Even if the integral curves of V and W appear to be coordinate curves, their parametrisation is not that of a coordinate system.

7.7 Geometric interpretation of the commutator

Let $p_0 \in M$, and let V, $W \in \mathfrak{X}(M)$ be linearly independent in a vicinity of p_0. Say we begin at p_0 and move a distance ϵ in the direction of W_{p_0} (i.e. along the integral curve of W which passes through p_0); we stop when we arrive at a point p_1. We do the same in the direction of V_{p_0} and arrive at a point p_2. Starting at p_1, we then move a distance ϵ in the direction of V_{p_1} and reach a point p_3. Finally, we start at p_2 and move a distance ϵ in the direction of W_{p_2}, reaching a point p_4 (see figure 7.2). Are p_3 and p_4 the same point? We can find out by evaluating an arbitrary function at both points and compare the results.

Let $f : M \longrightarrow \mathbb{R}$ be an arbitrary function.

$$f(p_4) - f(p_3) = f(p_4) - f(p_2) + f(p_2) - f(p_0) + f(p_0) - f(p_1) + f(p_1) - f(p_3)$$

$$= \epsilon\, W_{p_2}(f) + \epsilon\, V_{p_0}(f) - \epsilon\, W_{p_0}(f) - \epsilon\, V_{p_1}(f) + \mathcal{O}(\epsilon^2)$$

$$= \epsilon\left[W_{p_2}(f) - W_{p_0}(f) \right] - \epsilon\left[V_{p_1}(f) - V_{p_0}(f) \right] + \mathcal{O}(\epsilon^2)$$

$$= \epsilon^2\, V_{p_0}\!\left(W_{p_0}(f) \right) - \epsilon^2\, W_{p_0}\!\left(V_{p_0}(f) \right) + \mathcal{O}(\epsilon^3)$$

$$= \epsilon^2 [V, W]_{p_0}(f) + \mathcal{O}(\epsilon^3)$$

so

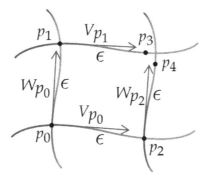

Figure 7.2. Geometric interpretation of the commutator. Moving the same distance in different directions along the integral curves of the coordinate basis vectors does not reach the same point, unless the commutator of the fields vanishes.

$$[V, W]_{p_0}(f) = \frac{f(p_4) - f(p_3)}{\epsilon^2} - \mathcal{O}(\epsilon)$$

We see that, as $\epsilon \to 0$, the following is true for all $f : M \longrightarrow \mathbb{R}$ such that $f \in C^2$:

$$[V, W]_{p_0}(f) = 0 \Leftrightarrow f(p_4) = f(p_3)$$

7.8 Exercise (polar coordinates)

Consider the 'unit' basis for polar coordinates on the Euclidean plane:

$$\hat{r} = \cos\theta \, \frac{\partial}{\partial x} + \sin\theta \, \frac{\partial}{\partial y}$$

$$\hat{\theta} = -\sin\theta \, \frac{\partial}{\partial x} + \cos\theta \, \frac{\partial}{\partial y}$$

Is this a coordinate basis?

Solution. Observe that $\theta = \arctan(\frac{y}{x})$ and thus

$$\frac{\partial\theta}{\partial x} = \frac{-\dfrac{y}{x^2}}{1 + \dfrac{y^2}{x^2}} = -\frac{y}{r^2} = -\frac{\sin\theta}{r}$$

$$\frac{\partial\theta}{\partial y} = \frac{\dfrac{1}{x}}{1 + \dfrac{y^2}{x^2}} = \frac{x}{r^2} = \frac{\cos\theta}{r}$$

Therefore,

$$[\hat{r}, \hat{\theta}] = \left[\cos\theta\,\frac{\partial}{\partial x} + \sin\theta\,\frac{\partial}{\partial y}, -\sin\theta\,\frac{\partial}{\partial x} + \cos\theta\,\frac{\partial}{\partial y}\right]$$

$$= -\cos\theta\left(\frac{\partial}{\partial x}\sin\theta\right)\frac{\partial}{\partial x} + \cos\theta\left(\frac{\partial}{\partial x}\cos\theta\right)\frac{\partial}{\partial y} - \sin\theta\left(\frac{\partial}{\partial y}\sin\theta\right)\frac{\partial}{\partial x}$$

$$+ \sin\theta\left(\frac{\partial}{\partial y}\cos\theta\right)\frac{\partial}{\partial y} + \sin\theta\left(\frac{\partial}{\partial x}\cos\theta\right)\frac{\partial}{\partial x} + \sin\theta\left(\frac{\partial}{\partial x}\sin\theta\right)\frac{\partial}{\partial y}$$

$$- \cos\theta\left(\frac{\partial}{\partial y}\cos\theta\right)\frac{\partial}{\partial x} - \cos\theta\left(\frac{\partial}{\partial y}\sin\theta\right)\frac{\partial}{\partial y}$$

$$= \left(\frac{\cos^2\theta\,\sin\theta}{r} - \frac{\cos^2\theta\,\sin\theta}{r} + \frac{\sin^3\theta}{r} + \frac{\cos^2\theta\,\sin\theta}{r}\right)\frac{\partial}{\partial x}$$

$$+ \left(\frac{\cos\theta\,\sin^2\theta}{r} - \frac{\cos\theta\,\sin^2\theta}{r} - \frac{\cos\theta\,\sin^2\theta}{r} - \frac{\cos^3\theta}{r}\right)\frac{\partial}{\partial y}$$

$$= \frac{\sin\theta}{r}\frac{\partial}{\partial x} - \frac{\cos\theta}{r}\frac{\partial}{\partial y} = -\frac{1}{r}\hat{\theta} \neq 0$$

We see that $\{\hat{r}, \hat{\theta}\}$ is not a coordinate basis. Let us see in detail why.
We know that $x = r\cos\theta$ and $y = r\sin\theta$, so

$$\frac{\partial x}{\partial\theta} = -r\sin\theta, \qquad \frac{\partial y}{\partial\theta} = r\cos\theta$$

In other words,

$$\frac{\partial}{\partial\theta} = -r\sin\theta\,\frac{\partial}{\partial x} + r\cos\theta\,\frac{\partial}{\partial y} \tag{7.7}$$

which means that $\hat{\theta} = \frac{1}{r}\frac{\partial}{\partial\theta}$.

Taking $e_r = \frac{\partial}{\partial r}$, $e_\theta = \frac{\partial}{\partial\theta}$, quadrilaterals are closed (see figure 7.3, left) and $[e_r, e_\theta] = 0$. But then e_θ grows linearly with r. To be able to 'measure' in the same way everywhere, it is customary to normalise e_θ by $e_\theta \longmapsto \hat{e}_\theta = \frac{1}{r}e_\theta$ $(=\hat{\theta})$, but then quadrilaterals are no longer closed (see figure 7.3, right) and we have $[e_r, \hat{e}_\theta] \neq 0$.

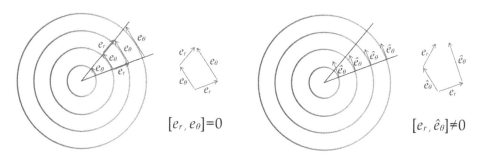

$[e_r, e_\theta] = 0$ $[e_r, \hat{e}_\theta] \neq 0$

Figure 7.3. The commutator for different bases in polar coordinates.

7.9 Exercise (gradient in terms of different bases)

(i) Find the dual bases of the coordinate basis $e_r = \frac{\partial}{\partial r}$, $e_\theta = \frac{\partial}{\partial \theta}$ and the non-coordinate basis $\hat{e}_r = \frac{\partial}{\partial r}$, $\hat{e}_\theta = \frac{1}{r}\frac{\partial}{\partial \theta}$.

(ii) Write the gradient of a function $f : M \longrightarrow \mathbb{R}$ in terms of the coordinate and non-coordinate bases.

Solution.

(i) For the coordinate basis $e_r = \frac{\partial}{\partial r}$, $e_\theta = \frac{\partial}{\partial \theta}$, we have

$$\frac{\partial}{\partial r}r = 1$$

$$\frac{\partial}{\partial \theta}r = 0$$

and

$$\frac{\partial}{\partial r}\theta = 0$$

$$\frac{\partial}{\partial \theta}\theta = 1$$

and therefore $\omega_r = dr$ and $\omega_\theta = d\theta$. For the non-coordinate basis $\hat{e}_r = \frac{\partial}{\partial r}$, $\hat{e}_\theta = \frac{1}{r}\frac{\partial}{\partial \theta}$, we have

$$\frac{\partial}{\partial r}r = 1$$

$$\frac{1}{r}\frac{\partial}{\partial \theta}r = 0$$

and therefore $\hat{\omega}_r = dr$. Setting $\hat{\omega}_\theta = a\,d\theta$, we obtain

$$\hat{\omega}_\theta(\hat{e}_\theta) = a\,d\theta(\hat{e}_\theta) = a\frac{1}{r}\frac{\partial\theta}{\partial\theta} = a\frac{1}{r}$$

and therefore $a = r$, that is, $\hat{\omega}_\theta = r\,d\theta$.

(ii) In the coordinate basis,

$$df(e_r) = e_r(f) = \frac{\partial}{\partial r}f$$

$$df(e_\theta) = e_\theta(f) = \frac{\partial}{\partial \theta}f$$

Therefore,

$$df = \frac{\partial f}{\partial r}\,dr + \frac{\partial f}{\partial \theta}\,d\theta$$

In the non-coordinate basis,

$$df = \hat{e}_r(f)\hat{\omega}_r + \hat{e}_\theta(f)\hat{\omega}_\theta = \frac{\partial f}{\partial r}\,\hat{\omega}_r + \frac{1}{r}\,\frac{\partial f}{\partial \theta}\,\hat{\omega}_\theta = \frac{\partial f}{\partial r}\,dr + \frac{\partial f}{\partial \theta}\,d\theta$$

Note that the 1-form df does not depend on the coordinates; its expression in terms of a given basis is what changes.

Chapter 8

Differential forms and exterior calculus

We have seen that the set of vector fields $\mathfrak{X}(M)$ on a manifold M has the structure of a vector space, where the sum and function–multiplication operations are defined pointwise. In the same way, the set of 1-forms $\mathfrak{X}^*(M)$ on M also has vector-space structure; it is the dual space of $\mathfrak{X}$. Suppose x^1, ..., x^n are local coordinates on an open set $U \subset M$; then, since $\{\frac{\partial}{\partial x^i}\}_{i=1, \ldots, n}$ is a basis of $\mathfrak{X}(M)$ and $\{dx^i\}_{i=1, \ldots, n}$ is a basis of $\mathfrak{X}^*(M)$, for any $X \in \mathfrak{X}(M)$ and any $\omega \in \mathfrak{X}^*(M)$ there exist scalar functions $\{\alpha_i \colon U \longrightarrow \mathbb{R}\}$, $\{\beta_i \colon U \longrightarrow \mathbb{R}\}$ ($i \in \{1, \ldots, n\}$) such that

$$X\,|_U = \sum_{i=1}^n \alpha_i\, \frac{\partial}{\partial x^i}$$

$$\omega\,|_U = \sum_{i=1}^n \beta_i\, dx^i$$

In this chapter we will study a special kind of tensor field called a 'differential form'. Differential forms generalise the concept of 1-form and other mathematical structures. They constitute a powerful tool, as we shall see in this chapter and others, because they simplify calculations, provide intuitive and graphical representations of relations and physical objects, tell us about the orientability of a manifold, etc. The study of differential forms is called 'exterior calculus'.

8.1 Definition (differential k-form)

We denote by $\Lambda^k(M)$ ($k \in \mathbb{N}$) the set of antisymmetric, k-linear functions on $\mathfrak{X}^*(M)$. Put differently, $\lambda \colon \underbrace{\mathfrak{X}(M) \times \cdots \times \mathfrak{X}(M)}_{k \text{ times}} \longrightarrow \mathbb{R} \in \Lambda^k(M)$ if and only if the following is true for all $v, v_1, \ldots, v_n \in \mathfrak{X}(M), a \in \mathbb{R}, j \in \{1, \ldots, k\}$:

(i) $\lambda(v_1, \ldots, v_{j-1}, v_j + v, v_{j+1}, \ldots, v_k) =$
$\lambda(v_1, \ldots, v_{j-1}, v_j, v_{j+1}, \ldots, v_k) + \lambda(v_1, \ldots, v_{j-i}, v, v_{j+1}, \ldots, v_k),$

(ii) $\lambda(v_1, \ldots, v_{j-1}, a\, v_j, v_{j+1}, \ldots, v_k) = a\, \lambda(v_1, \ldots, v_{j-1}, v_j, v_{j+1}, \ldots, v_k),$

(iii) $\lambda(v_{\pi(1)}, \ldots, v_{\pi(k)}) = (-1)^\pi \lambda(v_1, \ldots, v_k),$

where π is an element of S_k, the group of permutations of k elements, and

$$(-1)^\pi = \begin{cases} 1 & \text{if } \pi \text{ is an even permutation} \\ -1 & \text{if } \pi \text{ is an odd permutation} \end{cases}$$

Obviously, $\Lambda^k(M)$ has vector-space structure. Its elements are called 'differential k-forms'.

Let V, W be vector spaces over a field K. Then the Cartesian product $V \times W$ has the structure of a vector space over K:

$$(v_1, w_1) + (v_2, w_2) = (v_1 + v_2, w_1 + w_2)$$
$$a(v, w) = (a\, v, a\, w)$$

If $V \cap W = \{0\}$, this vector space is denoted by $V \oplus W$ and called the 'direct sum of V and W'. Note that $V \oplus \{0\}$ is a subset of $V \oplus W$ which is isomorphic to V and is therefore identified with V. The same is true for $W \oplus \{0\}$. Furthermore,

$$\dim(V \oplus W) = \dim(V) + \dim(W)$$

8.2 Differential forms and Grassmann algebra

We define

$$\mathscr{G}(M) \overset{\text{def}}{=} \bigoplus_{k=0}^{n} \Lambda^k(M)$$

$\mathscr{G}$ allows a 'natural' product $\wedge: \mathscr{G} \times \mathscr{G} \longrightarrow \mathscr{G}$ which is defined as follows. Given $\lambda \in \Lambda^k(M)$, $\mu \in \Lambda^l(M)$, the product $\lambda \wedge \mu$ is given by

$$\lambda \wedge \mu\,(v_1, \ldots, v_{k+l}) \overset{\text{def}}{=} \frac{1}{(k+l)!} \sum_{\pi \in S_{k+l}} (-1)^\pi \lambda(v_1, \ldots, v_k) \mu(v_{k+1}, \ldots, v_{k+l}) \in \Lambda^{k+l}(M)$$

The product $\wedge$, called 'exterior product' or sometimes 'wedge product', extends to all of $\mathscr{G}$ by requesting that it is distributive (with respect to vector addition) and associative. This gives $\mathscr{G}(M)$ the structure of an algebra. Note that $\wedge$ is not commutative: if $\lambda \in \Lambda^k(M)$ and $\mu \in \Lambda^l(M)$, then

$$\lambda \wedge \mu = (-1)^{kl} \mu \wedge \lambda \tag{8.1}$$

For a given $p \in M$, we use the notation $\Lambda_p^k(M) \overset{\text{def}}{=} \Lambda^k(M)|_p$, $\mathscr{G}_p(M) \overset{\text{def}}{=} \mathscr{G}(M)|_p$.

If $\{\omega_1, \ldots \omega_n\}$ is a basis of $\mathfrak{X}_p^*(M)$, then $\{\omega_{i_1} \wedge \cdots \wedge \omega_{i_k} \mid 1 \leqslant i_1 < i_2 < \cdots < i_k \leqslant n\}$ is a basis of $\Lambda_p^k(M)$ and

$$\bigcup_{k=0}^{n} \{\omega_{i_1} \wedge \cdots \wedge \omega_{i_k} \mid 1 \leqslant i_1 < i_2 < \cdots < i_k \leqslant n\}$$

is a basis of $\mathscr{G}_p(M)$. Therefore,

$$\dim\left(\Lambda_p^k(M)\right) = \binom{n}{k}$$

$$\dim\left(\mathscr{G}_p(M)\right) = \sum_{k=0}^{n}\binom{n}{k} = 2^n$$

$\mathscr{G}(M)$ is called the 'Grassmann algebra' of M, and its elements are called 'differential forms' on M.

8.3 Theorem (exterior derivative)

There exists a unique linear map $d: \mathscr{G}(M) \longrightarrow \mathscr{G}(M)$, called 'exterior derivative', with the following properties:

    (i) $d : \Lambda^k(M) \longrightarrow \Lambda^{k+1}(M)$.
    (ii) If $f \in \Lambda^0(M)$ (that is, $f : M \longrightarrow \mathbb{R}$), then $d(f) = df$ (i.e. d is the ordinary differential of f).
    (iii) If $\lambda \in \Lambda^k(M)$ and $\mu \in \mathscr{G}(M)$, then $d(\lambda \wedge \mu) = d\lambda \wedge \mu + (-1)^k \lambda \wedge d\mu$.
    (iv) $d^2 = 0$.

Proof. We will prove uniqueness first, assuming existence, and then we will prove the latter. This is because the way in which we should define the map d becomes evident when we prove uniqueness.

Uniqueness:

Suppose a linear map $d : \mathscr{G}(M) \longrightarrow \mathscr{G}(M)$ satisfying the conditions (*i*) to (*iv*) above exists. Let $U \subset M$ and let $\{x^1, \ldots, x^n\}$ be a local coordinate system in U. Furthermore, let $\omega \in \Lambda^k(M)$.

Since $\{dx^1, \ldots, dx^n\}$ is a basis of $\mathfrak{X}^*(U)$, then $\{dx^{i_1} \wedge \cdots \wedge dx^{i_k} \mid 1 \leqslant i_1 < \cdots < i_k \leqslant n\}$ is a basis of $\Lambda^k(U)$. Therefore, there exist $a_{i_1, \ldots, i_k} : U \longrightarrow \mathbb{R}$ (with $1 \leqslant i_1 < \cdots < i_k \leqslant n$) such that

$$\omega\big|_U = \sum a_{i_1, \ldots, i_k} \, dx^{i_1} \wedge \cdots \wedge dx^{i_k}$$

Note that the right-hand side of this equation is not a differential form on M, since it is only defined in U. Therefore, d cannot operate on it.

Let $U' \subset M$ such that $\bar{U}'$ is compact and $\bar{U}' \subset U$, and let $f : M \longrightarrow \mathbb{R}$ be a function such that $f|_{U'} \equiv 1$ and $f|_{M\setminus U} \equiv 0$. Finally, let $\omega' = \sum f \, a_{i_1, \ldots, i_k} \, d(f \; x^{i_1}) \wedge \cdots \wedge d(f \; x^{i_k})$. Then

$$\omega' \in \Lambda^k(M) \qquad \text{and} \qquad \omega'\big|_{U'} = \omega\big|_{U'}$$

Define $\mu = \omega' - \omega$. Note that $\mu|_{U'} \equiv 0$. Let $x_0 \in U'$, and let $g: M \longrightarrow \mathbb{R}$ be a function such that $g(x_0) = 1$ and $g|_{M\setminus U'} \equiv 0$. Then $g\mu \equiv 0$ on all of M. Since d is linear, $0 = d(g\,\mu) = (dg) \wedge \mu + g\,d\mu$. Evaluating this at x_0, we get $d\mu(x_0) = 0$. But x_0 is arbitrary, so $d\mu|_{U'} = 0$. In other words,

$$0 = d(\omega' - \omega)|_{U'} = (d\omega' - d\omega)|_{U'}$$

and, therefore,

$$d\omega'|_{U'} = d\omega|_{U'}$$

But

$$d\omega' = \sum d[f\,a_{i_1,\,\ldots,\,i_k}\,d(f\,x^{i_1}) \wedge \cdots \wedge d(f\,x^{i_k})]$$
$$\stackrel{(iii)}{=} \sum d(f\,a_{i_1,\,\ldots,\,i_k}) \wedge d(f\,x^{i_1}) \wedge \cdots \wedge d(f\,x^{i_k})$$
$$+ \sum f\,a_{i_1,\,\ldots,\,i_k}\,d[d(f\,x^{i_1}) \wedge \cdots \wedge d(f\,x^{i_k})]$$
$$\stackrel{(iv)}{=} \sum d(f\,a_{i_1,\,\ldots,\,i_k}) \wedge d(f\,x^{i_1}) \wedge \cdots \wedge d(f\,x^{i_k})$$

Since $f|_{U'} \equiv 1$ and $d\omega'|_{U'} = d\omega|_{U'}$, we have

$$d\omega|_{U'} = \sum_{i_1 < \ldots < i_k}\,\sum_{j=1}^{n} \frac{\partial}{\partial x^j}(a_{i_1,\,\ldots,\,i_k})dx^j \wedge dx^{i_1} \wedge \cdots \wedge dx^{i_k} \qquad (8.2)$$

Therefore, if d exists, it is given in terms of k-forms by equation (8.2) in U'. But U and U' are arbitrary, and all differential forms are sums of k-forms with $k \in \{0, \ldots, n\}$. Thus, d is unique (in U).

Existence;

We shall first define d locally.

Let $\{x^1, \ldots, x^n\}$ be a local coordinate system in an open set $U \subset M$. We define $d_U : \mathscr{G}(U) \longrightarrow \mathscr{G}(U)$ as follows. Given

$$\omega = \sum a_{i_1,\,\ldots,\,i_k}\,dx^{i_1} \wedge \cdots \wedge dx^{i_k} \;\in \Lambda^k(U)$$

define

$$d_U\omega = \sum \sum_{j=1}^{n} \frac{\partial}{\partial x^j}(a_{i_1,\,\ldots,\,i_k})dx^j \wedge dx^{i_1} \wedge \cdots \wedge dx^{i_k} \qquad (8.3)$$

and extend it to all $\mathscr{G}(U)$ by linearity.

Because of the way we built d_U, it satisfies (i) and (ii). Since d_U is linear and the exterior product is distributive with respect to vector addition, we need only prove that (iii) and (iv) hold on $\Lambda^k(U)$ for a single component of each form, as follows.

Let $\lambda = a_{i_1,\ldots,i_k}\,dx^{i_1} \wedge \cdots \wedge dx^{i_k}$ and $\mu = b_{j_1,\ldots,j_l}\,dx^{j_1} \wedge \cdots \wedge dx^{j_l}$. Then

$$d_U(\lambda \wedge \mu) = d_U(a_{i_1, ..., i_k} \, b_{j_1, ..., j_l} \, dx^{i_1} \wedge \cdots \wedge dx^{i_k} \wedge dx^{j_1} \wedge \cdots \wedge dx^{j_l})$$

$$= \sum_{r=1}^{n} \left[\frac{\partial}{\partial x^r}(a_{i_1, ..., i_k}) b_{j_1, ..., j_l} + \frac{\partial}{\partial x^r}(b_{j_1, ..., j_l}) a_{i_1, ..., i_k} \right]$$
$$\times dx^r \wedge dx^{i_1} \wedge \cdots \wedge dx^{i_k} \wedge dx^{j_1} \wedge \cdots \wedge dx^{j_l}$$

$$= \left(\sum_{r=1}^{n} \frac{\partial}{\partial x^r}(a_{i_1, ..., i_k}) dx^r \wedge dx^{i_1} \wedge \cdots \wedge dx^{i_k} \right)$$
$$\wedge \left(b_{j_1, ..., j_l} \, dx^{j_1} \wedge \cdots \wedge dx^{j_l} \right)$$
$$+ (-1)^k (a_{i_1, ..., i_k} \, dx^{i_1} \wedge \cdots \wedge dx^{i_k})$$
$$\wedge \left(\sum_{r=1}^{n} \frac{\partial}{\partial x^r}(b_{j_1, ..., j_l}) dx^r \wedge dx^{j_1} \wedge \cdots \wedge dx^{j_l} \right)$$

$$= (d_U\lambda) \wedge \mu + (-1)^k \lambda \wedge d_U\mu$$

$$(8.4)$$

so d_U satisfies (iii).

Now,

$$d^2{}_U\lambda = d_U \left(\sum_{r=1}^{n} \frac{\partial}{\partial x^r}(a_{i_1, ..., i_k}) dx^r \wedge dx^{i_1} \wedge \cdots \wedge dx^{i_k} \right)$$

$$= \sum_{r,s=1}^{n} \frac{\partial}{\partial x^s} \left(\frac{\partial}{\partial x^r}(a_{i_1, ..., i_k}) \right) dx^s \wedge dx^r \wedge dx^{i_1} \wedge \cdots \wedge dx^{i_k}$$

Since $dx^r \wedge dx^r = 0$, the terms where $s = r$ equal zero. For $s \neq r$, we have

$$\frac{\partial}{\partial x^s} \frac{\partial}{\partial x^r}(a_{i_1, ..., i_k}) dx^s \wedge dx^r = -\frac{\partial}{\partial x^r} \frac{\partial}{\partial x^s}(a_{i_1, ..., i_k}) dx^r \wedge dx^s$$

so all terms cancel by pairs. In other words, $d^2{}_U = 0$, i.e. d_U satisfies (iv).

Finally, we expand d to all of M as follows.

From the uniqueness result, any linear operator in $\mathscr{G}(U)$ that satisfies (i)–(iv) must be given by equation (8.3). In particular, if $U' \subset U$ then $\{x^1, ..., x^n\}|_{U'}$ is a local coordinate system for U' and

$$d_{U'}(\omega|_{U'}) = [d_U(\omega|_U)]\big|_{U'}$$

for all $\omega \in \mathscr{G}$. This allows us to define d globally: $\forall \omega \in \mathscr{G}$, $\forall U \subset M$, we write

$$(d\omega)\big|_U \overset{\text{def}}{=} d_U(\omega|_U) \qquad (8.5)$$

Note that d is well defined, for if $U, V \subset M$ are such that $U \cap V \neq \emptyset$, then

$$[d_U(\omega|_U)]\big|_{U \cap V} = d_{U \cap V}(\omega|_{U \cap V}) = [d_V(\omega|_V)]\big|_{U \cap V}$$

$\square$

8.4 Definition (volume element)

A 'volume element' of M is a choice of basis for $\Lambda^n(M)$, where $n = \dim(M)$.

Given that $\Lambda^n(M)$ is one-dimensional, a volume element of M is given by any nonzero element of $\Lambda^n(M)$.

8.5 Example

Let $\{dx^1, \ldots, dx^n\}$ be a basis of $\mathfrak{X}^*(M)$. Then $dx^1 \wedge \cdots \wedge dx^n$ is a volume element of M.

Since $\dim(\Lambda^n(M)) = 1$, a choice of volume element ω on M determines a natural isomorphism

$$\Lambda^n(M) \cong \mathbb{R}$$
$$r\,\omega \leftrightarrow r$$

8.6 Remark (vector calculus)

Given a volume element $\omega \in \Lambda^n(M)$ on M, we will build a function

$$m : \Lambda^{n-1}(M) \longrightarrow \mathfrak{X}(M) \cong \mathfrak{X}^{**}(M)$$

as follows.

If $\varphi \in \Lambda^{n-1}(M)$ and $\psi \in \Lambda^1(M) = \mathfrak{X}^*(M)$, then $\varphi \wedge \psi \in \Lambda^n(M)$. Therefore, $\exists\, a \in \mathbb{R}$ such that $\varphi \wedge \psi = a\,\omega$. This allows us to define $m(\varphi)[\psi] = a$.

Note that $m(\varphi) : \mathfrak{X}^*(M) \longrightarrow \mathbb{R}$, that is, $m(\varphi) \in \mathfrak{X}^{**}(M)$. But $\mathfrak{X}^{**}(M) \simeq \mathfrak{X}(M)$, so we may think of $m(\varphi)$ as a vector.

Now, an 'inner product' $\langle\,,\,\rangle$ on $\mathfrak{X}(M)$ defines an isomorphism

$$g : \mathfrak{X}(M) \longrightarrow \mathfrak{X}^*(M)$$
$$X \longmapsto g(X)$$

given by $g(X)[Y] = \langle X, Y \rangle$.

If $\{e_1, \ldots, e_n\}$ is a basis of $\mathfrak{X}(M)$ and $\{\psi^1, \ldots, \psi^n\}$ is its dual basis on $\mathfrak{X}^*(M)$, then

$$g(e_i) = g_{ij}\,\psi^j \tag{8.6}$$

where $g_{ij} = \langle e_i, e_j \rangle$. In particular, if $\{e_1, \ldots, e_n\}$ is orthonormal then

$$g_{ij} = \delta_{ij} \quad \text{and} \quad g(e_i) = \delta_{ij}\,\psi^j$$

For $M = \mathbb{R}^3$ ($\mathfrak{X} \simeq \mathbb{R}^3$, $\mathfrak{X}^* \simeq \mathbb{R}^3$), we define the following:
(i) Given $f : \mathbb{R}^3 \longrightarrow \mathbb{R} \in C^\infty$,

$$\nabla f \overset{\text{def}}{=} \text{grad}(f) \overset{\text{def}}{=} g^{-1}(df) \tag{8.7}$$

With respect to the usual coordinates in $\mathbb{R}^3$, we have

$$\nabla f = g^{-1}(df) = g^{-1}\left(\frac{\partial f}{\partial x^i}\, dx^i\right) = \delta^{ij}\, \frac{\partial f}{\partial x^i}\, \frac{\partial}{\partial x^j}$$

that is, its components are $(\frac{\partial f}{\partial x^1}, \frac{\partial f}{\partial x^2}, \frac{\partial f}{\partial x^3})$.

(ii) Given $V \in \mathfrak{X}(\mathbb{R}^3)$, we have $g(V) \in \Lambda^1(\mathbb{R}^3)$ and therefore

$$d(g(V)) \in \Lambda^2(\mathbb{R}^3) = \Lambda^{n-1}(\mathbb{R}^3)$$

Furthermore, since $m : \Lambda^{n-1}(\mathbb{R}^3) \longrightarrow \mathfrak{X}(\mathbb{R}^3)$ we have $m(d(g(V))) \in \mathfrak{X}(\mathbb{R}^3)$. Therefore, we may define

$$\nabla \times V \stackrel{\text{def}}{=} \text{rot}(V) \stackrel{\text{def}}{=} (m \circ d \circ g)(V) \qquad (8.8)$$

Now, since

$$g(V) = V_1\, dx^1 + V_2\, dx^2 + V_3\, dx^3 \in \Lambda^1(\mathbb{R}^3)$$

and $d^2 = 0$, we have

$$d(g(V)) = \frac{\partial V_1}{\partial x^2}\, dx^2 \wedge dx^1 + \frac{\partial V_1}{\partial x^3}\, dx^3 \wedge dx^1$$

$$+ \frac{\partial V_2}{\partial x^1}\, dx^1 \wedge dx^2 + \frac{\partial V_2}{\partial x^3}\, dx^3 \wedge dx^2$$

$$+ \frac{\partial V_3}{\partial x^1}\, dx^1 \wedge dx^3 + \frac{\partial V_3}{\partial x^2}\, dx^2 \wedge dx^3$$

$$= \left(\frac{\partial V_3}{\partial x^2} - \frac{\partial V_2}{\partial x^3}\right) dx^2 \wedge dx^3 + \left(\frac{\partial V_1}{\partial x^3} - \frac{\partial V_3}{\partial x^1}\right) dx^3 \wedge dx^1$$

$$+ \left(\frac{\partial V_2}{\partial x^1} - \frac{\partial V_1}{\partial x^2}\right) dx^1 \wedge dx^2$$

Therefore, the components of $\nabla \times V$ with respect to the usual coordinates in $\mathbb{R}^3$ are given by

$$(m \circ d \circ g)(V) = \left(\frac{\partial V_3}{\partial x^2} - \frac{\partial V_2}{\partial x^3}\right)\frac{\partial}{\partial x^1} + \left(\frac{\partial V_1}{\partial x^3} - \frac{\partial V_3}{\partial x^1}\right)\frac{\partial}{\partial x^2}$$

$$+ \left(\frac{\partial V_2}{\partial x^1} - \frac{\partial V_1}{\partial x^2}\right)\frac{\partial}{\partial x^3}$$

(iii) Given $V_1, V_2 \in \mathfrak{X}(\mathbb{R}^3)$, we have $g(V_1), g(V_2) \in \Lambda^1(\mathbb{R}^3)$. Therefore,

$$g(V_1) \wedge g(V_2) \in \Lambda^2(\mathbb{R}^3) = \Lambda^{n-1}(\mathbb{R}^3)$$

and the image of $g(V_1) \wedge g(V_2)$ under m is a vector: the 'vector product' of V_1 and V_2, given by

$$V_1 \times V_2 \overset{\text{def}}{=} m(g(V_1) \wedge g(V_2)) \tag{8.9}$$

(iv) Given $V \in \mathfrak{X}(\mathbb{R}^3)$, we have $m^{-1}(V) \in \Lambda^{n-1}(\mathbb{R}^3)$, and therefore $d \circ m^{-1}(V) \in \Lambda^n(\mathbb{R}^3)$ is a multiple of the volume element $\omega \in \Lambda^n(\mathbb{R}^3)$. This multiple is, save perhaps for its sign, the 'divergence of V'. In $\mathbb{R}^n$,

$$(-1)^{n-1} d[m^{-1}(V)] = (\nabla \cdot V)\omega \tag{8.10}$$

(i) and (iv) can be generalised to $M = \mathbb{R}^n$. (ii) and (iii), on the other hand, are specific to $M = \mathbb{R}^3$.

8.7 Lemma (vector calculus II)

    (i) $\nabla \times \nabla f = 0$ for arbitrary $f : \mathbb{R}^3 \longrightarrow \mathbb{R}^3$.
    (ii) $\nabla \cdot \nabla \times V = 0$ for arbitrary $V \in \mathfrak{X}(\mathbb{R}^3)$.

Proof. The two identities follow immediately from the above:
    (i) $\nabla \times \nabla f = m \circ d \circ g(g^{-1} \circ df) = m(d^2 f) = 0$.
    (ii) $d \circ m^{-1}(\nabla \times V) = d \circ m^{-1}(m \circ d \circ g(V)) = d^2(g(V)) = 0$.

                                                                     $\square$

8.8 Definition (closed and exact forms)

Let $\omega \in \mathscr{G}(M)$.
    (i) ω is called a 'closed form' if $d\omega = 0$.
    (ii) ω is called an 'exact form' if it is the differential of another form, that is, if $\exists \lambda \in \mathscr{G}$ such that $\omega = d\lambda$.

8.9 Corollary

All exact forms are closed (because $d^2 = 0$).
    The question of whether a closed form is exact gives rise to algebraic topology.

8.10 Orientation

Given an n-dimensional vector space V, $\Lambda^n(V)$ is one-dimensional. Therefore, $\Lambda^n(V) \cong \mathbb{R}$ (vector-space isomorphism). This implies that $\Lambda^n(V) \backslash \{0\}$ is disconnected—it is the union of two connected components.
    An 'orientation' in V is a choice of one of these two components. Therefore, V has two possible orientations: an ordered basis $\{\lambda^1, \ldots, \lambda^n\}$ of V^* determines an orientation, which is the component of $\Lambda^n(V)$ which contains $\lambda^1 \wedge \ldots \wedge \lambda^n$. Given two ordered bases $\{\lambda^1, \ldots, \lambda^n\}$ and $\{\lambda'^1, \ldots, \lambda'^n\}$, we have $\lambda'^i = A^i{}_j \lambda^j$ and

$\lambda'^1 \wedge \cdots \wedge \lambda'^n = \det(A^i{}_j)\lambda^1 \wedge \cdots \wedge \lambda^n$. Therefore, two ordered bases determine the same orientation if and only if the determinant of the basis-change matrix is positive.

8.11 Definition (orientable manifold)

Let (M, Φ) be an n-dimensional differentiable manifold. Then M is 'orientable' if and only if $\exists \mathcal{U}' \subset \Phi$ such that
  (i) $\mathcal{U}'$ is a cover for M.
  (ii) Given (φ, U), $(\psi, V) \in \mathcal{U}'$ with coordinates $(x^1, \ldots, x^n)$ and $(y^1, \ldots, y^n)$, respectively, the function $a : U \cap V \longrightarrow \mathbb{R}$ defined by

$$dx^1 \wedge \cdots \wedge dx^n = a \, dy^1 \wedge \cdots \wedge dy^n$$

  is positive in all of $U \cap V$.

An 'orientation' of M is a choice of $\mathcal{U}'$.

The function a defined above is the determinant of the Jacobian of $\varphi \circ \psi^{-1}$, that is, $a = \det(\frac{\partial}{\partial y^i} x^i) = \det(d(\varphi \circ \psi^{-1}))$. Therefore, a manifold that is orientable and connected has exactly two orientations, as mentioned above.

We summarise these results formally in the following theorems.

8.12 Theorem (orientability)

Let (M, Φ) be an n-dimensional differentiable manifold. If M has a volume element (that is, if there is an n-form ω on M which is never zero), then M is orientable.

8.13 Theorem (orientability II)

Let M be a connected n-dimensional differentiable manifold. Let

$$\mathcal{O} = \bigcup_{p \in M} \{\text{zero element in } \Lambda^n_p(M)\} \subset \Lambda^n(M)$$

Then one of the following is true:
  (i) $\Lambda^n(M) \backslash \mathcal{O}$ is connected, in which case M is not orientable.
  (ii) $\Lambda^n(M) \backslash \mathcal{O}$ has exactly two connected components, in which case M is orientable.

8.14 Remark (analytical mechanics)

Let $M = \mathbb{R}^3$. The movement of a particle with mass m on M under the influence of a force field F is given by

$$m \frac{d^2 x^i(t)}{dt^2} = F^i$$

Defining $p_i = m(\frac{dx^j}{dt})\delta_{ij}$, we obtain

$$\frac{dx^i}{dt} = \delta^{ij}\frac{p_j}{m}, \quad \frac{dp_j}{dt} = \delta_{ij}F^i$$

which are differential equations in $(x^1, x^2, x^3, p_1, p_2, p_3)$, the coordinate functions of $\mathfrak{X}^*(M)$. Therefore, the particle's orbit is merely the projection onto $\mathbb{R}^3$ of an integral curve on $\mathfrak{X}^*(M)$. In fact, the cotangent bundle $\mathfrak{X}^*(M)$ is the natural domain for the study of analytical mechanics on a manifold.

We will come back to this in chapter 16, where we will state the bases of Newtonian mechanics in the language that we have been constructing, and see how this structure allows for a greater mathematical simplicity and a more fundamental understanding.

IOP Publishing

Differential Topology and Geometry with Applications to Physics

Eduardo Nahmad-Achar

Chapter 9

Maps between manifolds

In this chapter, we shall define differential maps from one manifold to another and use these maps to carry structures (vectors, differentiable forms, tensors, functions) defined on one manifold to the other.

9.1 Definition (map between manifolds)

Let M and N be differentiable manifolds. Then $h : M \longrightarrow N$ is called a 'class-C^k map' if and only if $f {\circ} h \in C^k$ for all $f : N \longrightarrow \mathbb{R} \in C^k$.

Note that if $h : M \longrightarrow N$ is a class-C^k map, where $k \geqslant 1$, then h maps a curve $\alpha \in C^k$ on M to a curve $\beta \in C^k$ on N. This allows us to map vectors tangent to M to vectors tangent to N: given a point $p \in \alpha(t)$, we can associate the vector $\dot{\alpha}_p$ tangent to α at p to the vector $\dot{\beta}_{h(p)}$ tangent to β at $h(p)$, as shown in figure 9.1.

9.2 Definition (differential map)

Let $h : M \longrightarrow N$ be a class-C^k map with $k \geqslant 1$. Then the induced linear map $h_* : T_p(M) \longrightarrow T_{h(p)}(N)$ described above is called the 'differential of h'.

9.3 Lemma

(i) For all $f : N \longrightarrow \mathbb{R} \in C^k$,

$$(h_* X_p)f = X_p(f \circ h) \tag{9.1}$$

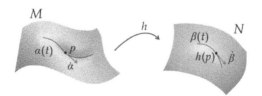

Figure 9.1. Mapping curves and tangent vectors from one manifold to another.

(ii) h_* is linear.

Proof. To prove (i), let $\alpha(t)$ be a curve such that $\alpha(0) = p$, $\dot\alpha(0) = X_p$. By definition,

$$
\begin{aligned}
(h_*)f &= (h \circ \alpha)^{\cdot}_{h_p}(f) \\
&= [f \circ (h \circ \alpha)]^{\cdot}_{(0)} \\
&= [(f \circ h) \circ \alpha]^{\cdot}_{(0)} \\
&= X_p(f \circ h)
\end{aligned}
\tag{9.2}
$$

To prove (ii), we write

$$
h_*(X_p + Y_p)[f] \overset{(9.2)}{=} (X_p + Y_p)[f \circ h] = X_p[f \circ h] + Y_p[f \circ h] \overset{(9.2)}{=} (h_*X_p)f + (h_*Y_p)f
$$

$$
h_*(aX_p) \overset{(9.2)}{=} (aX_p)[f \circ h] = a\big(X_p[f \circ h]\big) \overset{(9.2)}{=} a(h_*X_p)f
$$

$\square$

9.4 Definition (pull-back)

Let $h : M \longrightarrow N$ and $h_* T_p(M) \longrightarrow T_{h(p)}(N)$ be as above. Then, for $\omega \in T^*_{h(p)}(N)$ and $X_p \in T_p(M)$, the map

$$
h^* : T^*_{h(p)}(N) \longrightarrow T^*_p(M)
$$

defined by

$$
h^*\omega(X_p) \longmapsto \omega(h_*X_p)
$$

is called the 'pull-back' of h.

The pull-back of h, h^*, can be extended to rank-$\begin{pmatrix} 0 \\ s \end{pmatrix}$ tensors:

Given $T \in \underbrace{T^*_{h(p)}(N) \times \cdots \times T^*_{h(p)}(N)}_{s \text{ times}}$,

$$
h^* : \underbrace{T^*_{h(p)}(N) \times \cdots \times T^*_{h(p)}(N)}_{s \text{ times}} \longrightarrow \underbrace{T^*_p(M) \times \cdots \times T^*_p(M)}_{s \text{ times}}
$$

is such that

$$
(h^*T)(X^1_p, \ldots, X^s_p) = T(h_*X^1_p, \ldots, h_*X^s_p)
$$

When $s = 0$, we simply have functions from N to $\mathbb{R}$ and from M to $\mathbb{R}$ (see figure 9.2):

$$
\begin{aligned}
h^* : C^k(N, \mathbb{R}) &\longrightarrow C^k(M, \mathbb{R}) \\
f &\longmapsto f \circ h
\end{aligned}
$$

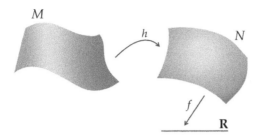

Figure 9.2. The pull-back acting on functions is just the usual composition of functions.

9.5 Example

Let $M = \mathbb{R}^3$ with coordinates (x, y, z), $N = \mathbb{R}^2$ with coordinates (x, y), and

$$h: M \longrightarrow N$$
$$(x, y, z) \longmapsto (x, y)$$

that is, h is the projection on the xy plane. Then for all $f: N \longrightarrow \mathbb{R}$

$$f \circ h(x, y, z) = f(x, y)$$

or, equivalently,

$$(h^*f)(x, y, z) = f(x, y)$$

Let $X_p \in T_p(M)$ with components (X, Y, Z), and let $g: M \longrightarrow \mathbb{R}$. Then

$$X_p[g] = \left(X\frac{\partial g}{\partial x} + Y\frac{\partial g}{\partial y} + Z\frac{\partial g}{\partial z} \right)_p$$

Therefore,

$$(h_*X_p)f = X_p[f \circ h] = \left(X\frac{\partial f}{\partial x} + Y\frac{\partial f}{\partial y} \right)_p$$

that is, $h_*X_p \in T_{h(p)}(N)$ has components (X, Y).

Let $\omega \in T^*_{h(p)}(N)$ with components (ω_1, ω_2); in other words, $\omega = \omega_1\, dx + \omega_2\, dy$. Then

$$h^*\omega(X_p) = \omega(h_*X_p) = (\omega_1\, dx + \omega_2\, dy)\left(X\frac{\partial}{\partial x} + Y\frac{\partial}{\partial y} \right) = \omega_1\, X + \omega_2\, Y$$

Thus, $h^*\omega \in T^*_p(M)$ has components $(\omega_1, \omega_2, 0)$: we write $h^*\omega = \omega_1\, dx + \omega_2\, dy + 0\, dz$.

9.6 Exercise

Repeat example 9.5 for the case in which $M = \mathbb{R}^2$, $N = \mathbb{R}^3$.

Solution.

$$h: M \longrightarrow N$$
$$(x, y) \longmapsto (x, y, 0)$$

Given $f: N \longrightarrow \mathbb{R}$, we have

$$(h^*f)(x, y) = f \circ h(x, y) = f(x, y, 0)$$

Given $X_p \in T_p(M)$ with components (X, Y) and $g: M \longrightarrow \mathbb{R}$,

$$X_p[g] = \left(X \frac{\partial g}{\partial x} + Y \frac{\partial g}{\partial y} \right)_p$$

Therefore,

$$(h_* X_p)f = X_p(f \circ h) = \left(X \frac{\partial (f \circ h)}{\partial x} + Y \frac{\partial (f \circ h)}{\partial y} \right)_p = \left(X \frac{\partial f(x, y, 0)}{\partial x} + Y \frac{\partial f(x, y, 0)}{\partial y} \right)_p$$

that is, $h_* X_p \in T_{h(p)}(N)$ has components $(X, Y, 0)$.

Let $\omega \in T^*_{h(p)}(N)$ with components $(\omega_1, \omega_2, \omega_3)$. Then

$$(h^*\omega)(X_p) = \omega(h_* X_p) = (\omega_1 \, dx, \omega_2 \, dy, \omega_3 \, dz)\left(X \frac{\partial}{\partial x}, Y \frac{\partial}{\partial y}, 0 \frac{\partial}{\partial z} \right) = \omega_1 X + \omega_2 Y$$

Therefore, $h^* \in T^*_p(M)$ has components (ω_1, ω_2): we write $h^*\omega = \omega_1 \, dx + \omega_2 \, dy$.

9.7 Example (vector field on a Möbius strip)

Consider the Möbius strip given by

$$h: [0, 2\pi] \times (-1, 1) \longrightarrow \mathbb{R}^3$$
$$(\theta, t) \longmapsto (x^1, x^2, x^3)$$

with

$$x^1 = 2 \cos \theta + t \cos \frac{\theta}{2} \cos \theta$$

$$x^2 = 2 \sin \theta + t \cos \frac{\theta}{2} \sin \theta$$

$$x^3 = t \sin \frac{\theta}{2}$$

that is, h maps the manifold $[0, 2\pi] \times (-1, 1)$ to a Möbius strip immersed in $\mathbb{R}^3$.

We take $v = \begin{pmatrix} 0 \\ 1 \end{pmatrix} = \frac{\partial}{\partial t} \in T([0, 2\pi] \times (-1, 1))$. Then, for any arbitrary function $f: \text{Möbius} \longrightarrow \mathbb{R}^3$, we have

$$h_*(v)f = \frac{\partial}{\partial t}(f \circ h) = \frac{\partial f}{\partial x^i} \frac{\partial x^i}{\partial t} = \cos \frac{\theta}{2} \cos \theta \frac{\partial f}{\partial x^1} + \cos \frac{\theta}{2} \sin \theta \frac{\partial f}{\partial x^2} + \sin \frac{\theta}{2} \frac{\partial f}{\partial x^3}$$

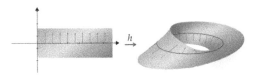

Figure 9.3. The differential map of the vector field $v = \frac{\partial}{\partial t}$ onto the Möbius strip is not continuous across the whole manifold, despite the fact that v and h are.

that is,

$$h_*(v) = \cos\frac{\theta}{2}\cos\theta\,\frac{\partial}{\partial x^1} + \cos\frac{\theta}{2}\sin\theta\,\frac{\partial}{\partial x^2} + \sin\frac{\theta}{2}\frac{\partial}{\partial x^3} \qquad (9.3)$$

and therefore

$$h_*(v)|_{\theta=0} = \frac{\partial}{\partial x^1}, \quad h_*(v)|_{\theta=2\pi} = -\frac{\partial}{\partial x^1} \qquad (9.4)$$

so that the vector field $h_*(v)$ is not continuous across the whole manifold despite the fact that v and h are (see figure 9.3).

9.8 Lie groups and Lie algebras

A 'Lie group' G is a differentiable manifold with group structure such that the group's operations are differentiable maps, that is, the map

$$h: G \times G \longrightarrow G$$
$$(x, y) \longmapsto xy^{-1}$$

is C^∞. (Note that this is a short and elegant way of saying that both the product and the inverse, precisely the operations one has in the group structure, are differentiable functions.)

For all $a \in G$ we define

$$L_a : G \longrightarrow G$$
$$x \longmapsto ax$$

the 'left translation' given by a.

It is evident that $L_a \in C^\infty$, as is its inverse function,

$$L_a^{-1} = L_{a^{-1}}: G \longrightarrow G$$
$$x \longmapsto a^{-1}x$$

Therefore, L_a is a diffeomorphism.

Let e be the identity element of G. For each tangent vector at e, $X_e \in T_e(G)$, we define the vector field

$$X: G \longrightarrow \mathfrak{X}(G)$$
$$g \longmapsto L_{g_*}X_e$$

$X(g)$ is a vector tangent at g: $X(g) \in T_g(G)$. For all $a \in G$, we have

$$L_{a*}(Xg) = L_{a*}L_{g*}(X_e) = L_{ag*}(X_e) = X(ag)$$

that is, $L_{a*}X = X$, so these vector fields are 'left-invariant'.

Let X and Y be left-invariant. Then

$$\begin{aligned}
L_{a*}((X + Y)g) &= L_{a*}(Xg + Yg) = L_{a*}L_{g*}(X_e + Y_e) \\
&= L_{ag*}(X_e + Y_e) = L_{ag*}(X_e) + L_{ag*}(Y_e) = X(ag) + Y(ag)
\end{aligned}$$

and, for $\alpha \in \mathbb{R}$,

$$L_{a*}(\alpha Xg) = L_{a*}L_{g*}(\alpha X_e) = L_{ag*}(\alpha X_e) = \alpha L_{ag*}(X_e) = \alpha X(ag)$$

Therefore, the set of left-invariant vector fields has vector-space structure. Since at each point $p \in G$ this space is a subset of $T_p(G)$, its dimension is less than or equal to n, the dimension of M. But we have defined a left-invariant vector field for each vector tangent at e; therefore, at each point the left-invariant vector fields form a vector space of the same dimension as M.

Once again, let X and Y be left-invariant. Let $\hat{g} : G \longrightarrow \mathbb{R}$, and let $q \in M$. Then

$$X(\hat{g} \circ L_a)(q) = X_q(\hat{g} \circ L_a) = L_{a*}(X_q)\hat{g}$$

that is,

$$X(\hat{g} \circ L_a) = L_{a*}X(\hat{g}) \tag{9.5}$$

Thus,

$$L_{a*}X(L_{a*}Y(\hat{g})) \overset{(9.5)}{=} X[(L_{a*}Y(\hat{g}))L_a] \overset{(9.5)}{=} X[Y(\hat{g} \circ L_a) \circ L_a]$$

Interchanging the roles of $L_{a*}X$ and $L_{a*}Y$ and those of X and Y, and subtracting, we obtain

$$[L_{a*}X, L_{a*}Y]\hat{g} = [[X, Y](\hat{g} \circ L_a)]L_a \overset{(9.5)}{=} [L_{a*}[X, Y]\hat{g}]L_a$$

that is, using the hypothesis,

$$[X, Y]\hat{g} = [L_{a*}X, L_{a*}Y]\hat{g} = L_{a*}[X, Y](\hat{g} \circ L_a)$$

Therefore, $[X, Y]$ is left-invariant.

The vector space of left-invariant vector fields, along with the operation $[\cdot, \cdot]$, forms an algebra A, the 'Lie algebra' associated to the Lie group G. The relationship between a Lie group and its associated Lie algebra is quite close, as the following results show (which we mention without proof):

    (i) Let G and G' be Lie groups, and let $\phi: G \longrightarrow G'$ be a group homomorphism. Then $\phi_* : A \longrightarrow A'$ is an algebra homomorphism between the algebras A and A' associated to G and G', respectively.

    (ii) Following the same notation as in (i), if $\varphi: A \longrightarrow A'$ is an algebra homomorphism and G is connected then there is a unique group homomorphism $\phi: G \longrightarrow G'$ such that $\varphi = \phi_*$.

The geometric interpretation of this relationship (see definition 13.5 and the following text) is that the Lie algebra A is the tangent space to the Lie group G at the identity point e:

$$A = T_e(G)$$

The study of Lie groups is important in mathematical physics; in particular, as will be seen in chapter 18, the group of a manifold's isometries is a Lie group. The association between a Lie group and its associated Lie algebra greatly facilitates their study (see definition 13.5).

IOP Publishing

Differential Topology and Geometry with Applications to Physics

Eduardo Nahmad-Achar

Chapter 10

Integration on manifolds

We have constructed differential calculus on a manifold using the way tangent vectors act on functions defined on the manifold. In this chapter we shall construct integral calculus for functions and differential forms on manifolds. In this way, we shall have transported the entire set of tools of calculus on $\mathbb{R}^n$ to manifolds of arbitrary dimension.

10.1 Remark

Let M be an n-dimensional orientable manifold. Then $\exists\, \mu \in \Lambda^n(M)$ such that $\mu \neq 0$ everywhere on M (therefore, μ is a volume element on M, see definition 8.4). Since μ is a basis of $\Lambda^n(M)$ (recall that $\dim[\Lambda^n(M)] = 1$), for all $\omega \in \Lambda^n(M)$ there is a function $f : M \longrightarrow \mathbb{R}$ such that $\omega = f\mu$.

10.2 Definition (integrability of functions and n-forms)

(i) $f : M \longrightarrow \mathbb{R}$ is 'integrable' if and only if the following is true:
   (a) f is bounded,
   (b) $\mathrm{supp}(f) = \{q \in M$ such that $f(q) \neq 0\}$ is compact,
   (c) f is continuous except in a set of measure zero.
(ii) $\omega \in \Lambda^n(M)$ is 'integrable' if and only if $\exists f : M \longrightarrow \mathbb{R}$ such that f is integrable and $\omega = f\mu$, where μ is a volume element as in remark 10.1.

Note that, for $\omega \in \Lambda^n(M)$ to be integrable, we do not require that $\omega \in C^\infty$, nor even that $\omega \in C^1$. The integrability of ω depends only on the integrability of f.

On the other hand, the integrability of ω does not depend on the volume element μ in M. If μ' is another volume element with the same orientation, then $\mu' = g\mu$ for some strictly positive $g \colon M \longrightarrow \mathbb{R} \in C^\infty$. Therefore, $f\mu = \frac{f}{g}\mu'$ and, if f is bounded, has compact support, and is continuous almost everywhere on M, the same will be true for $\frac{f}{g}$ ('almost everywhere' means 'everywhere except in a set of measure zero').

doi:10.1088/2053-2563/aadf65ch10

In a strict sense, we do not yet know what a 'set of measure zero' means because we do not yet have a way to measure sets. We could work only with continuous functions throughout this chapter and then extend our work to functions which are continuous except in a set of measure zero. In order to avoid repetition, however, we will deal with continuity almost everywhere from the start; the results obtained here can still be used for continuous functions whenever we lack a metric.

10.3 Definition (cube in a manifold)

Let (M, Φ) be an n-dimensional orientable differentiable manifold, $(U, \varphi) \in \Phi$, and $Q \subset U$. Then Q is called a 'cube in M' if and only if

$$\varphi(Q) = C = \{x \in \mathbb{R}^n \text{ such that } 0 \leqslant x^i \leqslant 1, i = 1, \ldots, n\}$$

We will denote Q without its boundary by

$$\mathring{Q} = \varphi^{-1}\{x \in \mathbb{R}^n \text{ such that } 0 < x^i < 1, i = 1, \ldots, n\}$$

10.4 Definition (integral of an n-form)

Let $\omega \in \Lambda^n(M)$ be integrable and such that $\text{supp}(\omega) \subset Q$, where Q is a cube in M. Let (U, φ) be a chart on M associated with Q, that is, $Q \subset U$. Since $\varphi^{-1*}(\omega) \in \Lambda^n(\mathbb{R}^n)$ and $\dim[\Lambda^n(\mathbb{R}^n)] = 1$, $\exists f : \mathbb{R}^n \longrightarrow \mathbb{R} \in C^0$ such that $\varphi^{-1*}(\omega) = f(x)dx^1 \wedge \cdots \wedge dx^n$. We then define

$$\int_M \omega \overset{\text{def}}{=} \int_C f \, dx^1 \cdots dx^n = \int_C f \, dv \tag{10.1}$$

where we have written $dv = dx^1 \cdots dx^n$ and $C = \varphi(Q)$.

10.5 Remark

To verify that equation (10.1) is well-defined, we must show that the integral's value is independent of the cube used.

Let Q' be another cube containing the support of ω and let (U', φ') be the chart associated with Q' with local coordinates $\{y^1, \ldots, y^n\}$ (that is, $\varphi'^{-1*}(\omega) = f'(y)dy^1 \wedge \cdots \wedge dy^n$ in U') and such that $Q' \subset U'$.

Let $J = \varphi' \circ \varphi^{-1}: \varphi(U \cap U') \longrightarrow \varphi'(U \cap U')$, and let ΔJ be the determinant of the Jacobian of this diffeomorphism. $\Delta J > 0$ if U and U' have the same orientation. Thus,

$$f(x) = f'(J(x))\Delta J \tag{10.2}$$

On the other hand, $\text{supp}(\omega) \subset Q$ and $\text{supp}(\omega) \subset Q'$. Therefore, $\text{supp}(\omega) \subset Q \cap Q'$ and

$$\int_{C'} f'(y)dv' = \int_{\varphi'(Q \cap Q')} f'(y)dv'$$

$$= \int_{\varphi(Q \cap Q')} f'[J(x)]|\Delta J|\, dv$$

$$= \int_{\varphi(Q \cap Q')} f(x)dv \qquad (10.3)$$

$$= \int_{C} f(x)dv$$

Therefore, $\int_{M} \omega$ is well-defined for all integrable n-forms whose support is contained in a cube in M.

In what follows, we will generalise the integral to arbitrary forms.

10.6 Definition (partition of unity)

A 'partition of unity' is a family of class-C^{∞} functions on M, which we will denote by $\{f_{\gamma}\}$, with the following properties:

(i) For all γ, $f_{\gamma} \geqslant 0$ on M.

(ii) $\{\mathrm{supp}(f_{\gamma})\}$ is a locally finite cover of M.

(iii) For all $p \in M$, $\sum_{\gamma} f_{\gamma}(p) = 1$.

The sum in (iii) is well-defined, since, by virtue of (ii), each $p \in M$ has a vicinity where only a finite number of f_{γ} are nonzero.

10.7 Definition (integration of general n-forms)

Let $\omega \in \Lambda^{n}(M)$ be integrable, with $K = \mathrm{supp}(\omega)$ compact. Let $Q_{1}, \dots, Q_{s}$ be cubes on M such that $\{\mathring{Q}_{1}, \dots, \mathring{Q}_{s}\}$ is a finite cover of K, and let $(U_{1}, \varphi_{1}), \dots, (U_{s}, \varphi_{s})$ be the coordinate charts associated to $Q_{1}, \dots, Q_{s}$. Finally, let $\{f_{i}\}$ be a partition of unity subordinate to the cover $\{\mathring{Q}_{1}, \dots, \mathring{Q}_{s}\}$, that is,

(i) for all $j \in \{1, \dots, s\}$, $\mathrm{supp}(f_{j}) \subset \mathring{Q}_{j}$;

(ii) for all $j > s$, $f_{j}|_{K} = 0$; and

(iii) $\sum_{j} f_{j} \equiv 1$.

We can then write $\omega = f_{1}\,\omega + \dots + f_{s}\,\omega$. We define

$$\int_{M} \omega \overset{\mathrm{def}}{=} \int_{M} f_{1}\,\omega + \dots + \int_{M} f_{s}\,\omega \qquad (10.4)$$

10.8 Theorem (properties of the integral)

Let M be an n-dimensional orientable differentiable manifold; μ be a volume element on M; ω, ω_1, $\omega_2 \in \Lambda^n(M)$ be integrable; and a_1, $a_2 \in \mathbb{R}$. We denote by $-M$ the manifold M with the opposite orientation. Then:

(i) $\displaystyle\int_{-M} \omega = -\int_M \omega$;

(ii) $\displaystyle\int_M (a_1\omega_1 + a_2\omega_2) = a_1 \int_M \omega_1 + a_2 \int_M \omega_2$;

(iii) if $\omega = f\,\mu$ with $f \geqslant 0$, then

$$\int_M \omega \geqslant 0$$

   where the equality is achieved if and only if $f = 0$ everywhere except in a set of measure zero;

(iv) if $F : M_1 \longrightarrow M_2$ is a diffeomorphism and $\bar\omega \in \Lambda^n(M_2)$ is integrable, then

$$\int_{M_1} F^*\bar\omega = \pm\int_{M_2} \bar\omega \tag{10.5}$$

   where the sign depends on whether F preserves orientation or not.

10.9 Remark

The most important result of integration on manifolds is Stokes' theorem, which relates the exterior differential operator d to the operator ∂. The result depends on 'manifolds with boundaries', which we shall introduce in what follows.

10.10 Definition (differentiable manifold with boundary)

We define $H^n \overset{\text{def}}{=} \{x = (x^1, \ldots, x^n) \in \mathbb{R}^n \text{ such that } x^n \geqslant 0\}$ with the induced topology of $\mathbb{R}^n$. We also define $\partial H^n \overset{\text{def}}{=} \{x \in H^n \text{ such that } x^n = 0\}$, which is homeomorphic to $\mathbb{R}^{n-1}$. Then:

(i) (M, Φ) is called an 'n-dimensional differentiable manifold with boundary, of class C^k' if and only if the following is true:
   (a) M is an n-dimensional Hausdorff space.
   (b) Φ is a collection of functions such that:
      (b1) $\{\text{dom}(\varphi) \text{ such that } \varphi \in \Phi\}$ is a cover of M;
      (b2) for all $\varphi \in \Phi$, φ: $\text{dom}(\varphi) \subset M \longrightarrow H^n$ is a homeomorphism;
      (b3) for all $\varphi, \psi \in \Phi$ such that $\text{dom}(\varphi) \cap \text{dom}(\psi) \neq \varnothing$,

      $$\psi \circ \varphi^{-1} : \varphi[\text{dom}(\varphi) \cap \text{dom}(\psi)] \subset H^n \longrightarrow H^n \in C^k$$

(ii) $\partial M \overset{\text{def}}{=} \{p \in M \text{ such that } \exists\, \varphi \in \Phi,\, \varphi(p) \in \partial H^n\}$ is called the 'boundary of M'.

From (i), if $\varphi(p) \in \partial H^n$ for a coordinate chart φ, then the same will be true for another coordinate chart ψ: we have $\psi(p) \in \partial H^n$. This is true because, if $U = \text{dom}(\varphi)$ and $V = \text{dom}(\psi)$, $\varphi(U)\backslash\partial H^n$ is an open set in $\mathbb{R}^n$ and therefore $\psi \circ \varphi^{-1}[\varphi(U) \cap \partial H^n] \subset \partial H^n$. Therefore, ∂M is well-defined.

10.11 Remark

The concepts of 'atlas', 'differentiable function' and so on, introduced earlier for manifolds, are equally translated to manifolds with boundaries.

10.12 Theorem (boundary orientability)

Let M be an orientable differentiable manifold with boundary $\partial M \neq \varnothing$. Then ∂M is orientable and an orientation on M determines an orientation for ∂M.

10.13 Theorem (Stokes' theorem)

Let M be an n-dimensional compact differentiable manifold, ∂M be the boundary of M with the induced orientation, and $\omega \in \Lambda^{n-1}(\partial M)$ be integrable. Then

$$\int_M d\omega = \int_{\partial M} \omega \tag{10.6}$$

Before proving Stokes' theorem, we comment on the orientation induced by the volume element $\mu \in M$ on ∂M.

Consider $M = C^2 = \{x = (x^1, x^2) \in \mathbb{R}^2 \text{ such that } 0 \leqslant x^i \leqslant 1, i = 1, 2\}$ with the orientation given by $\mu = dx^1 \wedge dx^2$, as shown in figure 10.1.

The orientation induced on ∂M is as shown in figure 10.2.

In other words, if we define

$$C^2_{(i, 0)} = C^2(x^i = 0)$$
$$C^2_{(i, 1)} = C^2(x^i = 1)$$

as the sides of $\partial M = \partial C^2$, we have

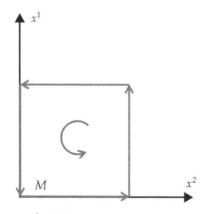

Figure 10.1. C^2 with the orientation given by $\mu = dx^1 \wedge dx^2$.

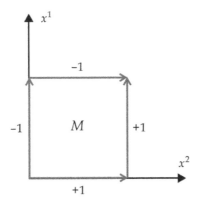

Figure 10.2. Induced orientation on the boundary, from the orientation of the manifold $M = C^2$.

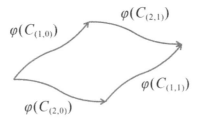

Figure 10.3. Induced orientation on the boundary for a general cube $Q \subset M$.

$$\partial C^2 = \sum_{i=1}^{2} \sum_{k=0}^{1} (-1)^{i+k} C^2_{(i,\,k)} \tag{10.7}$$
$$= C^2_{(1,\,1)} - C^2_{(2,\,1)} - C^2_{(1,\,0)} + C^2_{(2,\,0)}$$

In general, for a cube $Q \subset M$ with $\dim(M) = n$, we have

$$\partial Q = \varphi^{-1} \sum_{i=1}^{n} \sum_{k=0}^{1} (-1)^{i+k} \varphi(Q)_{(i,\,k)}$$

Figure 10.3 shows this schematically.

In terms of the volume element $\mu \in \Lambda^n(H^n)$, the orientation induced by μ on $\partial H^n = \{x \in \mathbb{R}^n \text{ such that } x^n = 0\}$ must be such that the 'positive orientation' of ∂H^n is given by a vector normal to H^n pointing 'outwards'. Writing

$$\begin{aligned}
\mu &= dx^1 \wedge \cdots \wedge dx^n \\
&= (-1)^{n-1} dx^n \wedge dx^1 \wedge \cdots \wedge dx^{n-1} \\
&= (-1)^n (-dx^n) \wedge dx^1 \wedge \cdots \wedge dx^{n-1} \\
&= (-1)^n (-dx^n) \wedge \partial \mu
\end{aligned} \tag{10.8}$$

where $\partial\mu \in \Lambda^{n-1}(H^n)$ is the natural volume element on ∂H^n and $-dx^n = d(-x^n)$ is the dual form of $-\frac{\partial}{\partial x^n}$ (i.e. the vector pointing 'outwards' on H^n), and noting that the face of the cube C^n intersecting ∂H^n is $C^n_{(n,0)}$, we see that the map

$$C^n_{(n,0)} \xrightarrow{(\varphi^{-1}, \varphi^{-1*})} (\partial M, \partial\mu)$$

preserves the orientation of ∂H^n for even n and inverts it for odd n.

Proof. By virtue of equation (10.4) (see definition 10.7), it suffices to prove the theorem for an ω whose support is contained in the interior $\mathring{Q}$ of a cube Q associated with a chart (U, φ).

Let $\omega \in \Lambda^{n-1}(M)$. Suppose that $\mathrm{supp}(\omega) \subset \mathring{Q}$. Let $\{x^1, \ldots, x^n\}$ be a local coordinate system on U. Then $\exists\, a_i : M \longrightarrow \mathbb{R}$ for $i \in \{1, \ldots, n\}$ such that

$$\varphi^{-1*}(\omega) = \sum_{i=1}^{n}(-1)^{i-1}a_i\, dx^1 \wedge \cdots \wedge dx^{i-1} \wedge dx^{i+1} \wedge \cdots \wedge dx^n \tag{10.9}$$

Therefore,

$$\varphi^{-1*}(d\omega) = d\varphi^{-1*}(\omega) = \sum_{i=1}^{n}\left(\frac{\partial a_i}{\partial x^i}\right)dx^1 \wedge \cdots \wedge dx^n$$

whereby

$$\int_M d\omega = \int_C \sum_{i=1}^{n}\left(\frac{\partial a_i}{\partial x^i}\right)dv = \sum_{i=1}^{n}\int_0^1 \cdots \int_0^1 \frac{\partial a_i}{\partial x^i}\, dx^1 \cdots dx^n$$

The ith term of the sum on the right-hand side of the previous equation is equal to

$$\int_0^1 \cdots \int_0^1 [a_i(x^1, \ldots, x^{i-1}, 1, x^{i+1}, \ldots, x^n) \tag{10.10}$$
$$- a_i(x^1, \ldots, x^{i-1}, 0, x^{i+1}, \ldots, x^n)]\, dx^1 \cdots \widehat{dx^i} \cdots dx^n$$

where we have denoted by $\widehat{dx^i}$ the differential which is omitted, having already integrated over the corresponding coordinate function.

We have the following cases:

(a) $Q \cap \partial M = \varnothing$. In this case,

$$\varphi(\mathring{Q}) = \{x = (x^1, \ldots, x^n) \in \mathbb{R}^n \text{ such that } 0 < x^i < 1, i = 1, \ldots, n\}$$

Because $\mathrm{supp}(\omega) \subset \mathring{Q}$, $a_i = 0$ for $x^i = 1$ or $x^i = 0$. Therefore, the integrand in equation (10.10) is zero for all $i \in \{1, \ldots, n\}$, whereby

$$\int_M d\omega = 0$$

But $\omega|_{\partial M} = 0$ because $\mathrm{supp}(\omega) \subset \mathring{Q}$, that is,

$$\int_{\partial M} \omega = 0$$

so trivially

$$\int_M d\omega = \int_{\partial M} \omega \qquad (10.6)$$

(b) $Q \cap \partial M \neq \emptyset$. In this case,

$$\varphi^{-1}(\mathring{Q}) = \{x \in \mathbb{R}^n \text{ such that } 0 < x^i < 1, \, i = 1, \dots, n-1, \, 0 \leqslant x^n < 1\}$$

so the integrand in equation (10.10) is zero for $i \neq n$, that is,

$$\int_M d\omega = -\int_0^1 \cdots \int_0^1 a_n(x^1, \dots, x^{n-1}, 0)dx^1 \cdots dx^{n-1}$$

Since $\text{supp}(\omega|_{\partial M}) \subset Q \cap \partial M$, we have, in terms of the local coordinates (see equation (10.9)),

$$\varphi^{-1*}(\omega|_{\partial M}) = (-1)^{n-1}a_n(x^1, \dots, x^{n-1}, 0)dx^1 \wedge \cdots \wedge dx^{n-1}$$

Therefore,

$$\int_{\partial M} \omega = (-1)^{n-1} \int_0^1 \cdots \int_0^1 a_n(x^1, \dots, x^{n-1}, 0)dx^1 \cdots dx^{n-1}$$

and thus

$$\int_M d\omega = (-1)^n \int_{\partial M} \omega = \int_{\partial M} \omega$$

where the last equality is for ∂M with the orientation induced by M.

□

Throughout the above proof, we have assumed that the exterior differentiation operator commutes with the pull-back, that is, that if $h : M \longrightarrow N$ then

$$h^* \circ d = d \circ h^*$$

This can be proven by induction over k for $\omega \in \Lambda^k(N)$:

- Let $a(x) \in \Lambda^0$ and $X_p \in T_p(M)$. Then

$$h^*(da)X_p = da(h_*X_p) = (h_*X_p)a = X_p(a \circ h) = X_p(h^*a) = d(H^*a)X_p$$

- Suppose h^* commutes with d for all $\omega \in \Lambda^l(N)$ with $l < k$. Let $\lambda \in \Lambda^k(N)$. Without loss of generality, we may write

$$\lambda = a(x)dx^{i_1} \wedge \cdots \wedge dx^{i_k}$$

We define

$$\omega_1 = a \, dx^{i_1}$$
$$\omega_2 = dx^{i_2} \wedge \cdots \wedge dx^{i_k}$$

Then $\lambda = \omega_1 \wedge \omega_2$ and $d\omega_2 = 0$, since $d^2 = 0$. Therefore,

$$
\begin{aligned}
d[h^*(\lambda)] &= d[h^*(\omega_1 \wedge \omega_2)] \\
&= d[(h^*\omega_1) \wedge (h^*\omega_2] \\
&= (dh^*\omega_1) \wedge (h^*\omega_2) - (h^*\omega_1) \wedge (dh^*\omega_2) \\
&= (dh^*\omega_1) \wedge (h^*\omega_2) - (h^*\omega_1) \wedge (h^*d\omega_2) \\
&= h^*(d\omega_1) \wedge (h^*\omega_2) \\
&= h^*(d\omega_1 \wedge \omega_2) \\
&= h^*d(\omega_1 \wedge \omega_2) \\
&= h^*(d\lambda)
\end{aligned}
\tag{10.11}
$$

10.14 Example (Gauss's theorem)

Let $M = V \subset \mathbb{R}^3$ be compact, and A be a vector field which, without loss of generality, we may write as

$$A = (a_{23} - a_{32})\mathbf{i} + (a_{31} + a_{13})\mathbf{j} + (a_{12} - a_{21})\mathbf{k}$$

Define the 2-form

$$\omega = (a_{23} - a_{32})dx^2 \wedge dx^3 + (a_{31} - a_{13})dx^3 \wedge dx^1 + (a_{12} - a_{21})dx^1 \wedge dx^2 \in \Lambda^2(M)$$

Then

$$d\omega = (\nabla \cdot A)dx^1 \wedge dx^2 \wedge dx^3$$

and Stokes' theorem tells us that

$$
\begin{aligned}
\int_V \nabla \cdot A \, dv &= \int_V (\nabla \cdot a)dx^1 \wedge dx^2 \wedge dx^3 \\
&= \int_M d\omega \overset{\text{Stokes}}{=} \int_{\partial M} \omega = \int_{\partial V} A \cdot ds
\end{aligned}
\tag{10.12}
$$

where ds is the area element on ∂V. Equation (10.12) is Gauss' theorem from advanced calculus on $\mathbb{R}^3$.

10.15 Example (Stokes' theorem in $\mathbb{R}^3$)

Let $M = V \subset \mathbb{R}^2$ be compact. We imagine $\mathbb{R}^2$ to be immersed in $\mathbb{R}^3$. Let

$$A = a_1 \, \mathbf{i} + a_2 \, \mathbf{j}$$

be a vector field in M and define the 1-form

$$\omega = \sum_i a_i \, dx^i = a_1 \, dx^1 + a_2 \, dx^2 \in \Lambda^1(M)$$

Then

$$d\omega = -\left(\frac{\partial a_1}{\partial x^2} - \frac{\partial a_2}{\partial x^1}\right) dx^1 \wedge dx^2$$

and

$$\nabla \times A = -\left(\frac{\partial a_1}{\partial x^2} - \frac{\partial a_2}{\partial x^1}\right)\mathbf{k}$$

and Stokes' theorem tells us that

$$\int_V (\nabla \times A) \cdot ds = \int_M d\omega \stackrel{\text{Stokes}}{=} \int_{\partial M} \omega = \int_{\partial V} A \cdot dl \qquad (10.13)$$

where ds is the area element on V and dl is the line element on ∂V. Equation (10.13) is Stokes' theorem from advanced calculus in $\mathbb{R}^3$. (If we do not imagine M to be immersed in $\mathbb{R}^3$, equation (10.13) is known as Green's theorem.)

10.16 Examples

(i) Let $M = \mathbb{S}^2$. Let μ be an exact form on M, that is, $\mu = d\alpha$ for some other form α. Then

$$\int_M \mu = \int_{\mathbb{S}^2} d\alpha = \int_{\partial \mathbb{S}^2} \alpha = 0$$

because $\partial \mathbb{S}^2 = \varnothing$. In other words, the integral of any exact form on $\mathbb{S}^2$ is zero.

(ii) Let $\omega = x^1 \, dx^2 \wedge dx^3$ be a 2-form on $\mathbb{R}^3$. Then $d\omega = dx^1 \wedge dx^2 \wedge dx^3$ is the natural volume element on $\mathbb{R}^3$. Therefore,

$$\int_{\mathbb{S}^2} \omega|_{\mathbb{S}^2} = \int_{\mathbb{B}^3} dx^1 \, dx^2 \, dx^3 = \frac{4}{3}\pi$$

where $\mathbb{B}^3$ is the unit ball in $\mathbb{R}^3$: $\partial \mathbb{B}^3 = \mathbb{S}^2$.

(iii) Let λ be a 2-form on $\mathbb{S}^2$. Then $d\lambda \in \Lambda^3(\mathbb{S}^2)$ and, therefore, $d\lambda = 0$. In other words, all 2-forms on $\mathbb{S}^2$ are closed. But $\omega|_{\mathbb{S}^2}$ has nonzero integral (cf. (ii)) and, therefore, is not exact (cf. (i)). Thus, not all closed forms are exact.

(iv) The harmonic oscillator in $M = \mathbb{R}^2$ is given by

$$\frac{d^2x}{dt^2} + \Omega^2 x = 0 \qquad (10.14)$$

To put this in the language of forms, we write

$$\frac{dx}{dt} = y$$

$$\frac{dy}{dt} = -\Omega^2 x \qquad (10.15)$$

Thus, equation (10.14) is satisfied along curves on which the quantities $\alpha \stackrel{\text{def}}{=} dx - y\,dt$ and $\beta \stackrel{\text{def}}{=} dy + \Omega^2 x\,dt$ are zero. We define

$$\gamma \stackrel{\text{def}}{=} \Omega^2 x \alpha + y\beta$$

Then

$$\gamma = d\left(\frac{1}{2}y^2 + \frac{1}{2}\Omega^2 x^2\right)$$

and $d\gamma = 0$. Therefore, along a curve l on which $\alpha = \beta = 0$ we have

$$d\left(\frac{1}{2}y^2 + \frac{1}{2}\Omega^2 x^2\right) = \gamma = \Omega^2 x \alpha + y\beta = 0$$

and

$$0 = \int_l d\left(\frac{1}{2}y^2 + \frac{1}{2}\Omega^2 x^2\right) = \left(\frac{1}{2}y^2 + \frac{1}{2}\Omega^2 x^2\right)\Big|_{p_1}^{p_2}$$

where p_1 and p_2 are the extrema of l. But $(\frac{1}{2}y^2 + \frac{1}{2}\Omega^2 x^2)$ is the oscillator's energy. Therefore, energy is conserved.

10.17 A look at cohomology and homology

We have two operators, d and ∂, which are related to each other by Stokes' theorem and have the properties $d^2 = 0$ and $\partial^2 = 0$. In particular, we have the 'co-chain'

$$\cdots \xrightarrow{d^{(k-2)}} \Lambda^{k-1}(M) \xrightarrow{d^{(k-1)}} \Lambda^k(M) \xrightarrow{d^{(k)}} \Lambda^{k+1}(M) \xrightarrow{d^{(k+1)}} \cdots$$

where $d^{(k)}$ denotes simply the exterior differentiation operator d and the index between parentheses, k, indicates that it acts on $\Lambda^k(M)$.

Because $d^2 = 0$, we know that the image of $d^{(k-1)}$ (that is, the set of exact k-forms) is contained in the kernel of $d^{(k)}$ (that is, the set of closed k-forms):

$$\text{img}[d^{(k-1)}] \subset \ker[d^{(k)}]$$

We can form the quotient

$$H^k(M) = \ker[d^{(k)}]\,/\,\text{img}[d^{(k-1)}]$$

(for $H^0(M)$, we take as the start of the co-chain $\{0\} \longrightarrow \Lambda^0(M)$ with the identity map which sends 0 to the function $f \equiv 0$). $H^k(M)$ has the structure of a group; it is the 'kth de Rham cohomology group'. As we have seen, not all closed forms are exact; Georges de Rham's cohomology theory is a classification of closed forms on a manifold: given two closed forms $\lambda, \mu \in \Lambda^k(M)$, we say that λ and μ are 'cohomologous' if and only if $\lambda - \mu$ is exact (that is, if they differ by an exact form or, in other words, if they are equivalent in $H^k(M)$).

Note that, if M has m connected components, then $H^0(M) \cong \mathbb{R}^m$, since any function $f: M \longrightarrow \mathbb{R} \in C^1$ such that $df = 0$ (i.e. any locally constant function) is constant over each connected component of M.

Intuitively, cohomology is a way to count 'holes' in a topological space. If we remove the origin of the real line, for instance, one point on one side of the line cannot be continuously moved onto a point on the other side of the line: we have a 'one-dimensional hole' between them that prevents it. These two points, one on each side, essentially constitute a 'zero-dimensional sphere', S^0. In the same way, if we remove a point or a disc from a plane, while any two points may be continuously moved one onto the other by taking any path which avoids the hole in the plane, a loop S^1 around the hole cannot be contracted to a point: we have a 'two-dimensional hole'. If we take a more complicated manifold such as a two-dimensional torus, T^2, it is easy to see that both an azimuthal and a polar loop around the torus may not be contracted down to a point, yet these loops surround different 'holes' in the manifold: the torus has then two two-dimensional holes. In $\mathbb{R}^3$ a 'three-dimensional hole' may be detected by a two-dimensional sphere S^2 surrounding it, which may not be contracted to a point, etc. (Note that some authors call 'n-dimensional hole' what we are here calling an $n + 1$-dimensional hole.)

The relationship to differential forms is that a hole has no boundary *and* it is not the boundary of something. Having no boundary immediately makes one think of closed loops and surfaces, and not being the boundary of something discards those closed loops and surfaces which bound something inside them. So, for example, a sphere S^n has no boundary but it is also the boundary of nothing: a hole.

Now, we have seen that a differential form is closed if its derivative vanishes, and it is exact if it is the derivative of another form. Being closed amounts to having no boundary, and being exact means that it bounds a piece of the manifold. Since a hole has no boundary and is not the boundary of something, we can then study them by studying differential forms which are closed but not exact, and this is what the theory of cohomology is about. The differential structure of the manifold gives us information of its geometric structure.

A typical example is to consider the 1-form

$$\mu = f_1 \, dx + f_2 \, dy$$

where the function $f = (f_1, f_2) : \mathbb{R}^2 \backslash \{0\} \to \mathbb{R}^2$ is defined as

$$f(x, y) = \left(\frac{-y^2}{x^2 + y^2}, \frac{x^2}{x^2 + y^2} \right)$$

We now ask if μ is exact, i.e. if there exists a function $F : U \subset M \to \mathbb{R}$ of class C^2 such that $\mu = dF$. This function, if it exists, would satisfy $(\partial F/\partial x) = f_1$ and $(\partial F/\partial y) = f_2$, and since $(\partial^2 F/\partial x \, \partial y) = (\partial^2 F/\partial y \, \partial x)$ we have

$$\frac{\partial f_1}{\partial y} = \frac{\partial f_2}{\partial x}$$

in other words, μ is a closed 1-form. However, integrating around a loop and using Stokes' theorem we would have

$$\int_0^{2\pi} dF = \int_{\partial S^1} F = 0$$

but this cannot be true since, using polar coordinates for simplicity,

$$\frac{d}{d\theta} F(\cos\theta, \sin\theta) = -f_1 \sin\theta + f_2 \cos\theta = 1$$

Therefore, F cannot exist, and μ is a closed but *not* exact form. The differential form μ as defined above is measuring that there exists a hole in the domain $\mathbb{R}^2 \setminus \{0\}$ where it is defined.

Seen in the complex domain μ takes the form

$$\mu = dz/z$$

and the fact that it is not exact is equivalent to the fact that the residue in a Laurent series is the coefficient of $1/z$, and this is what contributes to the integral along a closed path.

In general, we can define

$$\beta^k = \dim[H^k(M)]$$

The β^k ($k \in \{0, 1, ...\}$) are called 'Betti numbers' and are topological invariants of M. Intuitively, we have

$$\beta^0 = \text{number of connected components of } M$$
$$\beta^1 = \text{number of two−dimensional holes on } M$$
$$\beta^2 = \text{number of cavities (three−dimensional holes) on} M$$

etc.

A few examples are:

- $M = \mathbb{S}^1$ has $\beta^0 = 1$, $\beta^1 = 1$, $\beta^{j \geqslant 2} = 0$.
- $M = \mathbb{T}^2$ has $\beta^0 = 1$, $\beta^1 = 2$, $\beta^2 = 1$, $\beta^{j \geqslant 3} = 0$.
- $M = \text{double torus has } \beta^0 = 1$, $\beta^1 = 4$, $\beta^2 = 1$, $\beta^{j \geqslant 3} = 0$.

Stokes' theorem implies a duality between the de Rham cohomology and the 'homology of chains', which is given by

$$\cdots \xleftarrow{\partial^{(k-1)}} M^{k-1} \xleftarrow{\partial^{(k)}} M^k \xleftarrow{\partial^{(k+1)}} M^{k+1} \xleftarrow{\partial^{(k+2)}} \cdots$$

where $\partial^{(k)}$ is the boundary operator on M^k (a k-dimensional manifold). Similarly, since $\partial^2 = 0$, we have that

$$\mathrm{img}[\partial^{(k+1)}] \subset \ker[\partial^{(k)}]$$

and we can form the 'homology groups'

$$H_k = \ker[\partial^{(k)}] \, / \, \mathrm{img}[\partial^{(k+1)}]$$

The study of homology and cohomology groups and their properties corresponds to algebraic topology.

IOP Publishing

Differential Topology and Geometry with Applications to Physics

Eduardo Nahmad-Achar

Chapter 11

Integral curves and Lie derivatives

So far, we have been able to differentiate functions, but not vector fields or 1-form fields. The cause of this is that differentiation implies a comparison between the values at two points on the manifold and two tangent vectors (or 1-forms) at different points belong to different vector spaces. To compare them, we need to 'drag' one of them towards the other. The obtained object will depend on how we perform the drag. The 'Lie derivative' constitutes a 'natural' drag which we will construct and study in this chapter along with its geometric interpretation.

11.1 Definition (integral curve of a vector field)

Let X be a class-C^k vector field on M with $k \geqslant 1$, and let $\gamma : I \subset \mathbb{R} \longrightarrow M$ be a curve on M such that, for all $p \in \mathrm{img}(\gamma)$, $\dot{\gamma}(p) = X_p$. Then γ is called an 'integral curve of X'.

11.2 Theorem (uniqueness of integral curves)

Any vector field X on M defines a unique integral curve γ through each point $p \in M$ and for which $\gamma(0) = p$.

Proof. Let $p \in M$, $\{x^i\}$ be a local coordinate system, and x_p^i be the ith coordinate of p. A curve with parameter t has derivative $\frac{dx^i(t)}{dt}$, and the condition that this curve be an integral curve of X is

$$\frac{dx^i(t)}{dt} = X^i(t) \tag{11.1}$$

along with the initial condition $x^i(0) = x_p^i$. Since $X \in C^k$ with $k \geqslant 1$, the theorem of existence and uniqueness of solutions to ordinary differential equations guarantees that, for a sufficiently small interval of t, equation (11.1) with the aforementioned initial condition has a unique solution. $\qquad\square$

doi:10.1088/2053-2563/aadf65ch11

11.3 Lemma (coordinate system adapted to a vector field)

Let $p \in M$, and let X be a vector field on M such that $X_p \neq 0$. Then there is a neighbourhood U of p and a coordinate system $\{y^i\}$ in U such that $X = \frac{\partial}{\partial y^1}$ in U.

Proof. Let $n = \dim(M)$, U be a neighbourhood of p, $\Sigma \subset U$ be an $(n-1)$-dimensional hypersurface such that X is not tangent to Σ at any point (see figure 11.1), and $(y^2, ..., y^n)$ be the coordinates of Σ. Then, by virtue of theorem 11.2, given $p \in \Sigma$ there exists a unique integral curve γ_p of X which passes through p and satisfies $\gamma_p(0) = p$.

For $p \in \Sigma$, we define y^1 along γ_p as $y^1 = \int dt$ such that $y^1 = 0$ on Σ and we take $(y^2, ..., y^n)$ to be constant along γ_p. This way, $\{y^1\}_{i \in \{1, ..., n\}}$ is the coordinate system we seek. □

11.4 Definition (congruence and complete integral curve)

    (i) An integral curve $\gamma(t)$ on M is 'complete' if and only if $\gamma(t)$ is defined for all values of t.

    (ii) A set of integral curves is called a 'congruence'. (Note that any two curves of a congruence do not intersect each other, since they are integral curves of the same vector field.)

In what follows, we will consider that all integral curves are complete.

11.5 Definition (drag along curves)

Let X be a vector field on M, $p \in M$, and $\gamma(t)$ be an integral curve of X which passes through p and satisfies $\gamma(t_0) = p$. Then for all $s \in \mathbb{R}$ we define (see figure 11.2)

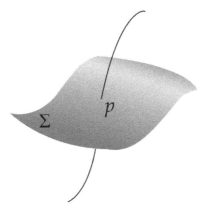

Figure 11.1. Integral curve of vector field X which passes through the point p. Σ is an $(n-1)$-dimensional hypersurface such that X is not tangent to it at any point, locally.

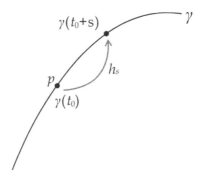

Figure 11.2. The function h_s, which allows us to drag functions (and in general tensor fields) along integral curves of a vector field.

$$h_s : M \longrightarrow M$$
$$p \longmapsto \gamma(t_0 + s)$$

By construction, $h_{t+s} = h_{s+t} = h_s \circ h_t$, h_0 is the identity function, and $h_{-s} = (h_s)^{-1}$. Therefore, $\{h_s\}_{s \in \mathbb{R}}$ is a one-parameter group. The transform group $\{h_s\}_{s \in \mathbb{R}}$ induced by X allows us to drag (or transport) functions, and in general tensor fields, along the integral curves of X, as we shall now see.

11.6 Definition (Lie derivative of a function)

Let X be a vector field on M, $p \in M$, $\gamma(t)$ be an integral curve of X passing through p, $\{h_t\}_{t \in \mathbb{R}}$ be the group of transformations induced by X, and $f : M \longrightarrow \mathbb{R} \in C^1$. Then

$$(\pounds_X f)_p \overset{\text{def}}{=} \lim_{t \to 0} \frac{f(h_t(p)) - f(p)}{t} \qquad (11.2)$$

is called the 'Lie derivative of f with respect to X' at p.

Note that definition 11.6 is independent of any coordinate system.

11.7 Theorem (on the Lie derivative of a function)

For all $f : M \longrightarrow \mathbb{R}$,

$$(\pounds_X f)_p = X_p f \qquad (11.3)$$

Proof. Using the coordinate system adapted to the integral curves of X as in lemma 11.3, we have

$$y^1(h_t(p)) = y^1(p) + t$$
$$y^j(h_t(p)) = y^j(p) \quad \forall j \geqslant 2$$

Thus,

$$(\pounds_X f)_p = \lim_{t \to 0} \frac{f(h_t(p)) - f(p)}{t} = \left(\frac{\partial f}{\partial y^1}\right)_p \overset{\text{Lemma 11.3}}{=} X_p f$$

whereby equation (11.3) is valid in any coordinate system. $\qquad\qquad$ $\square$

Recall that h_t induces a differential map $h_{t*} : T_{h_{-t(p)}}(M) \longrightarrow T_p(M)$. This can be used to compare vectors at different points.

11.8 Definition (Lie derivative of a vector field)

Let $p \in M$; X, Y be vector fields on M; γ be an integral curve of X passing through p; and $\{h_t\}_{t \in \mathbb{R}}$ be the group of transformations induced by X. Then

$$(\pounds_X Y)_p \overset{\text{def}}{=} \lim_{t \to 0} \frac{1}{t}\left[Y_p - h_{t*} Y_{h_{-t(p)}}\right] \qquad (11.4)$$

is called the 'Lie derivative of Y with respect to X' at p. A schematic diagram is shown in figure 11.3.

11.9 Geometric interpretation of the Lie derivative

We shall use the notation from definition 11.8. Let $\{\gamma_i\}$ be the congruence formed by the integral curves of X and $\{\alpha_i\}$ be the one formed by the integral curves of Y.

Let $q = h_{-t}(p)$. h_t maps the integral curve α_1 to the curve β (see figure 11.4). Therefore, h_{t*} maps $T_{\alpha_1}(M)$ to $T_\beta(M)$. The Lie derivative of Y along X compares $Y_p \in T_p(M)$ to $h_{t*} Y_q \in T_p(M)$ in the limit in which $t \to 0$.

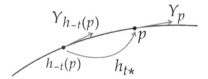

Figure 11.3. Dragging a vector field Y along integral curves of another vector field X.

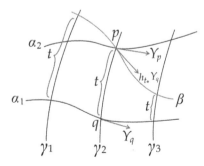

Figure 11.4. Geometric interpretation of the Lie derivative. See the text for details.

11.10 Theorem (on the Lie derivative of a vector field)

Let p, X and Y be as before. Then

$$(\pounds_X Y)_p = [X, Y]_p \tag{11.5}$$

Proof. We again use the coordinate system from lemma 11.3, which is adapted to the integral curves of X. Then

$$(\pounds_X Y)^i_p = \lim_{t \to 0} \frac{1}{t}\left[Y^i(y^1_p, y^\alpha_p) - Y^i(y^1_p - t, y^\alpha_p)\right] = \left(\frac{\partial Y^i}{\partial y^1}\right)_p = X_p(Y^i)$$

The rightmost term in this equation is not a vector (see equation (7.2)), but it would be one if we subtracted $Y_p(X^i)$ from it. Now, since $X^i = \delta^i_1$, we have $Y_p(X^i) = 0$. Thus,

$$(\pounds_X Y)^i_p = X_p(Y^i) - Y_p(X^i) = [X, Y]^i_p \tag{11.6}$$

Equation (11.6) is a vector equation and is thus valid in any coordinate system. $\square$

11.11 Corollary

In a local coordinate system in which $X = \frac{\partial}{\partial x^1}$, we have

$$(\pounds_X Y)^i_p = \frac{\partial}{\partial x^1} Y^i \tag{11.7}$$

Proof. Since $X = \frac{\partial}{\partial x^1}$, we have $X^j = \delta^j_1$, so

$$(\pounds_X Y)^i = [X, Y]^i = X^j \frac{\partial}{\partial x^j} Y^i - Y^j \frac{\partial}{\partial x^j} X^j = \frac{\partial}{\partial x^1} Y^i \tag{11.8}$$

$\square$

This corollary tells us that the Lie derivative is the coordinate-independent version of the partial derivative.

Looking at the definitions of the Lie derivative of functions and vector fields, it becomes clear that for 1-form fields we will use the pull-back induced by the transform group h_t.

11.12 Definition (Lie derivative of a 1-form field)

Let $X \in \mathfrak{X}(M)$, $p \in M$, γ be an integral curve of X passing through p, and ω be a 1-form field on M. Then

$$(\pounds_X\omega)_p \stackrel{\text{def}}{=} \lim_{t\to 0} \frac{\omega_p - h^*_{-t}\omega_{h_{-t}}(p)}{t} \tag{11.9}$$

A schematic diagram of equation (11.9) is shown in figure 11.5.

11.13 Lemma (calculation of the Lie derivative of a 1-form)

$\pounds_X\omega$ is a 1-form field whose components are

$$(\pounds_X\omega)_j = \omega_{j,\,i}X^i + \omega_k X^k{}_{,j} \tag{11.10}$$

Proof. Using Leibnitz's rule for an arbitrary $Y \in \mathfrak{X}(M)$,

$$\pounds_X(\omega(Y)) = (\pounds_X\omega)(Y) + \omega(\pounds_X Y)$$
$$= (\pounds_X\omega)(Y) + \omega[X,\ Y]$$
$$= (\pounds_X\omega)(Y) + \omega_k(Y^k{}_{,i}X^i - X^k{}_{,i}Y^i)$$

But $\omega(Y)$ is a real-valued function, so

$$\pounds_X(\omega(Y)) = X(\omega(Y)) = X^i\frac{\partial}{\partial x^i}(\omega_j Y^j) = \omega_{j,\,i}X^iY^j + \omega_j Y^j{}_{,i}X^i$$

Solving for $(\pounds_X\omega)(Y)$,

$$(\pounds_X\omega)(Y) = \omega_{j,\,i}X^iY^j + \omega_k X^k{}_{,j}Y^j \tag{11.11}$$

and, since Y is arbitrary,

$$(\pounds_X\omega)_j = \omega_{j,\,i}X^i + \omega^k X^k{}_{,j} \tag{11.12}$$

$\square$

11.14 Theorem (Lie derivative of an *n*-form)

Let M be an m-dimensional differentiable manifold, $X \in \mathfrak{X}(M)$ be any vector field on M, and $\omega \in \mathfrak{X}^{*n}(M)$ be an n-form on M. Then

$$\pounds_X\omega = d[\omega(X)] + d\omega(X) \tag{11.13}$$

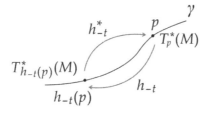

Figure 11.5. Dragging a 1-form field ω along integral curves of a vector field X.

Proof. We will prove this theorem by induction:
  (i) 'ω a 0-form' on M: that is, ω is a function $f : M \longrightarrow \mathbb{R}$. Then

$$\omega(X) = 0$$

and

$$d[\omega(X)] = 0$$

On the other hand,

$$d\omega(X) \overset{\text{def}}{=} df(X) = X[f] \overset{\text{theorem 11.7}}{=} \pounds_X f = \pounds_X \omega \tag{11.14}$$

  (ii) 'ω a 1-form': $\omega = \omega_i \, dx^i$. Then $\omega(X) = \omega_i X^i$, and thus

$$d[\omega(X)] = d(\omega_i X^i) = \omega_i X^i)_{,k} \, dx^k \tag{11.15}$$

and

$$d\omega = d(\omega_i \, dx^i) = \omega_{i,\,k} \, dx^k \wedge dx^i$$

Therefore,

$$d\omega(X) = \omega_{i,\,k}[dx^k(X)dx^i - dx^i(X)dx^k]$$
$$= \omega_{i,\,k} X^k \, dx^i - \omega_{i,\,k} X^i \, dx^k \tag{11.16}$$

Adding equations (11.15) and (11.16),

$$d[\omega(X)] + d\omega(X) \quad = \quad \omega_{i,\,k} X^k \, dx^i + \omega_i X^i_{\,,\,k} \, dx^k$$
$$\overset{i \leftrightarrow k}{=} \quad (\omega_{i,\,k} X^k + \omega_k X^k_{\,,\,i})dx^i \tag{11.17}$$
$$\overset{\text{Lemma 11.13}}{=} \pounds_X \omega$$

  (iii) '0,1, ..., n-forms $\Longrightarrow (n + 1)$-forms': any $(n + 1)$-form ω may be written as $\omega = \omega_{k_1 \cdots k_{n+1}} dx^{k_1} \wedge \cdots \wedge dx^{k_{n+1}}$ with $(k_1, \ldots, k_{n+1}) \in \Pi_{n+1}$, the group of permutations of $(n + 1)$-elements. It therefore suffices to prove that, if it holds for forms $\lambda \in \mathfrak{X}^{*n}(M)$ and $\mu \in \mathfrak{X}^*(M)$, it must hold for $\omega = f \lambda \wedge \mu$ with $f : M \longrightarrow \mathbb{R}$. On the one hand,

$$\pounds_X \omega \overset{\text{Leibnitz}}{=} (\pounds_X f)\lambda \wedge \mu + f(\pounds_X \lambda) \wedge \mu + f \lambda \wedge (\pounds_X \mu)$$

Using $X[f] = df(X)$ and the induction hypothesis, we have

$$\pounds_X \omega = df(X)\lambda \wedge \mu + f \, d[\lambda(X)] \wedge \mu + f[d\lambda(X)] \wedge \mu$$
$$+ f \lambda \wedge d[\mu(X)] + f \lambda \wedge (d\mu)(X) \tag{11.18}$$

On the other hand, using points (ii) and (iii) from theorem 8.3 and noting that $\lambda(X) \in \mathfrak{X}^{*n-1}(M)$, $df \in \mathfrak{X}^*(M)$ and $d\lambda \in \mathfrak{X}^{*n+1}(M)$, we have

$$d[\omega(X)] = d[f\,\lambda(X) \wedge \mu + (-1)^n f\,\lambda \wedge \mu(X)]$$
$$= df[\lambda(X) \wedge \mu] + f\,d[\lambda(X)] \wedge \mu + (-1)^n f\,\lambda(X) \wedge d\mu \qquad (11.19)$$
$$+ (-1)^n df[\lambda \wedge \mu(X)] + (-1)^n f\,d\lambda \wedge \mu(X) + f\,\lambda \wedge d[\mu(X)]$$

and

$$d\omega(X) = [df \wedge \lambda \wedge \mu - f\,d\lambda \wedge \mu + (-1)^n f\,\lambda \wedge d\mu](X)$$
$$= df(X)\lambda \wedge \mu - df \wedge [(\lambda \wedge \mu)(X)] + f\,d\lambda(X) \wedge \mu \qquad (11.20)$$
$$+ (-1)^{n+1} f\,d\lambda \wedge \mu(X) + (-1)^n f\,\lambda(X) \wedge d\mu + (-1)^n(-1)^n f\,\lambda \wedge d\mu(X)$$

Adding equations (11.19) and (11.20), we obtain equation (11.18). $\square$

11.15 Exercise (Lie derivative of a tensor)

Calculate the components of $\pounds_X T$ for a rank-$\binom{1}{1}$ tensor T.

Solution. Let Y be a vector field and ω be a 1-form field. Then $T(Y, \omega)$ is a function mapping M to $\mathbb{R}$ and

$$\pounds_X T(Y, \omega) = X[T(Y, \omega)]$$
$$= X^i(T^a{}_b Y^b \omega_a)_{,i} \qquad (11.21)$$
$$= X^i(T^a{}_{b,i} Y^b \omega_a + T^a{}_b Y^b{}_{,i} \omega_a + T^a{}_b Y^b \omega_{a,i})$$

On the other hand, using Leibnitz's rule,

$$\pounds_X T(Y, \omega) = (\pounds_X T)(Y, \omega) + T(\pounds_X Y, \omega) + T(Y, \pounds_X \omega)$$
$$= (\pounds_X T)(Y, \omega) + T^a{}_b X^i Y^b{}_{,i} \omega_a - T^a{}_c Y^b X^c{}_{,b} \omega_a \qquad (11.22)$$
$$+ T^a{}_b Y^b X^i \omega_{a,i} + T^c{}_b Y^b \omega_a X^a{}_{,c}$$

Putting equations (11.21) and (11.22) together, we obtain

$$(\pounds_X T)(Y, \omega) = T^a{}_{b,i} X^i Y^b \omega_a + T^a{}_c Y^b X^c{}_{,b} \omega_a - T^c{}_b Y^b \omega_a X^a{}_{,c}$$

that is,

$$(\pounds_X T)^a{}_b = T^a{}_{b,i} X^i + T^a{}_c X^c{}_{,b} - T^c{}_b X^a{}_{,c} \qquad (11.23)$$

11.16 Exercise (Lie derivative with respect to a commutator)

Let X, Y, and Z be vector fields on M. Prove that

$$\pounds_{[X,\,Y]} Z = \pounds_X \pounds_Y Z - \pounds_Y \pounds_X Z \qquad (11.24)$$

Solution. Using Jacobi's identity, we have

$$\pounds_{[X,\,Y]}Z = [[X,\,Y],\,Z]$$
$$= - [[Z,\,X],\,Y] - [[Y,\,Z],\,X]$$
$$= [Y,\,[Z,\,X]] + [X,\,[Y,\,Z]]$$
$$= - [Y,\,[X,\,Z]] + [X,\,[Y,\,Z]]$$
$$= \pounds_X\pounds_Y Z - \pounds_Y\pounds_X Z$$

11.17 Exercise (contraction and the Lie derivative)

Prove that the Lie derivative commutes with the contraction operator.

Solution. We know that

$$\pounds_X \delta^{\,a}_{\,b} = \delta^{\,a}_{\,b,\,c}X^c + X^c_{\,,b}\delta^{\,a}_{\,c} - X^a_{\,,c}\delta^{\,c}_{\,b} = X^a_{\,,b} - X^a_{\,,b} = 0$$

Therefore,

$$\pounds_X \delta^{\,b}_{\,a}T^a_{\,b} = (\pounds_X\delta^{\,b}_{\,a})T^a_{\,b} + \delta^{\,b}_{\,a}\,\pounds_X T^a_{\,b} = \delta^{\,b}_{\,a}\,\pounds_X T^a_{\,b} \qquad (11.25)$$

11.18 Exercise (Lie derivative in an arbitrary basis)

From theorem 11.10, we know that, in a coordinate basis, we have

$$(\pounds_X Y)^i = [X,\,Y]^i = X^j\frac{\partial}{\partial x^j}Y^i - Y^j\frac{\partial}{\partial x^i}X^i$$

Prove that, in an arbitrary basis $\{e_i\}$, we have

$$(\pounds_X Y)^i = X^j e_j(Y^i) - Y^j e_j(X^i) + X^j Y^k\big(\pounds_{e_j}e_k\big)^i \qquad (11.26)$$

Solution. We know that

$$\pounds_X Y = \pounds_X(Y^i e_i) = \pounds_X(Y^i)e_i + Y^i\pounds_X e_i$$

But

$$\pounds_X(Y^i) = X(Y^i) = X^j e_j(Y^i)$$

and, by the anticommutativity of the commutator,

$$\pounds_X e_i = -\pounds_{e_i}X = -\pounds_{e_i}(X^j e_j)$$

Substituting,

$$\pounds_X Y = X^j e_j(Y^i)e_i - Y^i\pounds_{e_i}(X^j e_j)$$
$$= \big[X^j e_j(Y^i) - Y^j e_j(X^i)\big]e_i - Y^i X^j\pounds_{e_i}e_j$$

Using the commutator's anticommutativity again in the last term,

$$\pounds_{e_i} e_j = -\pounds_{e_j} e_i$$

so, renaming the index i to k,

$$\pounds_X Y = \left[X^j e_j(Y^k) - Y^j e_j(X^k) \right] e_k + X^j Y^k \pounds_{e_j} e_k$$

Therefore,

$$(\pounds_X Y)^i = X^j e_j(Y^i) - Y^j e_j(X^i) + X^j Y^k \left(\pounds_{e_j} e_k \right)^i$$

IOP Publishing

Differential Topology and Geometry with Applications to Physics

Eduardo Nahmad-Achar

Chapter 12

Linear connections

We have introduced the concept of 'Lie transport' or 'Lie derivative', which allows us to differentiate vectors and, in general, tensors. But the Lie derivative is not general enough, since it depends on a choice of a vector field. Similarly, Lie transport is not what we could call 'parallel transport', as it depends on the chosen trajectory. In this chapter we shall introduce a new derivative, the 'covariant derivative', which not only compensates for the variation due to the manifold itself, but will also allow us to construct a parallel transport. In addition, it will allow us to define the 'curvature' and 'torsion' of a manifold and build its geodesics.

12.1 Definition (linear connection)

Let X, Y and Z be class-C^1 vectors on M, $f\colon M \longrightarrow \mathbb{R}$, $a \in \mathbb{R}$, and

$$\nabla\colon \mathfrak{X}(M) \times \mathfrak{X}(M) \longrightarrow \mathfrak{X}(M)$$
$$(X,\, Y) \longmapsto \nabla_X Y$$

be such that
- (i) $\nabla_X(aY + Z) = a\nabla_X Y + \nabla_X Z$.
- (ii) $\nabla_X(fY) = f\nabla_X Y + (X f)Y$.
- (iii) $\nabla_{fX+aY}(Z) = f\nabla_X Z + a\nabla_Y Z$.

Then ∇ is called a 'linear connection' on M and $\nabla(X,\, Y) \equiv \nabla_X Y$ is called the 'covariant derivative of Y with respect to X'.

∇ is not a tensor, since it is not linear in its second argument (cf. (ii)). ∇ is linear in its first argument, X (in contrast to the Lie derivative), so ∇Y is a rank-$\binom{1}{1}$ tensor. ∇Y is called the 'covariant derivative of Y'.

12.2 Definition (parallel transport)

Let X and Y be vector fields on M such that

$$\nabla_X Y = 0$$

Then Y is said to be 'parallel-transported' along X.

12.3 Definition (connection coefficients)

Let $\{e_a\}$ be a field of bases on $\mathfrak{X}(M)$. Since $\nabla_{e_a} e_b \in \mathfrak{X}(M)$, there exists $\Gamma^c{}_{ba} \in \mathbb{R}$ such that

$$\nabla_{e_a} e_b = \Gamma^c{}_{ba} e_c \tag{12.1}$$

The scalars $\Gamma^c{}_{ba}$ are called 'coefficients of the connection' ∇.

12.4 Remark (justification of definition 12.3)

$$
\begin{aligned}
\nabla_X Y &= \nabla_{(X^a e_a)}(Y^b e_b) \\
&= X^a \nabla_{e_a}(Y^b e_b) \\
&= X^a e_a(Y^b) e_b + X^a(\nabla_{e_a} e_b) Y^b \\
&= (X[Y^c] + \Gamma^c{}_{ba} X^a Y^b) e_c
\end{aligned}
\tag{12.2}
$$

that is, $\nabla_X Y$ is completely specified by the connection coefficients $\Gamma^c{}_{ba}$. If we define

$$Y^c{}_{;a} \overset{\text{def}}{=} e_a(Y^c) + \Gamma^c{}_{ba} Y^b \tag{12.3}$$

then, by equation (12.2),

$$(\nabla_X Y)^c = Y^c{}_{;a} X^a \tag{12.4}$$

that is, $Y^c{}_{;a}$ are the components of the rank-$\binom{1}{1}$ tensor ∇Y.

In terms of coordinates, equation (12.3) is written as

$$Y^c{}_{;a} = Y^c{}_{,a} + \Gamma^c{}_{ba} Y^b \tag{12.5}$$

Note that definition 12.1 does not completely fix the connection: we still have freedom to define it as long as our definition satisfies criteria (i), (ii), and (iii) from definition 12.1.

The first term from the right-hand side of equation (12.5) can always be calculated, but the second one ($\Gamma^c{}_{ba} Y^b$) depends on the chosen connection.

12.5 Example (motivation)

Let $M = \mathbb{R}^2$ with Cartesian coordinates $\{x^1, x^2\}$ and basis $\{e_1 = \frac{\partial}{\partial x^1}, e_2 = \frac{\partial}{\partial x^2}\}$.

We do not yet know how to build connections, but we know that e_1 is parallel-transported along e_2 and vice versa. Suppose, then, that we have a linear connection ∇ such that $\nabla_{e_i} e_j = 0$, whereby $\Gamma^k{}_{ij} = 0$.

Now let $M = \mathbb{R}^2$ with polar coordinates $\{r, \theta\}$ and basis $\{e_r = \frac{\partial}{\partial r}, e_\theta = \frac{1}{r}\frac{\partial}{\partial \theta}\}$. The fact that the vectors are not constant on M makes differentiation non-intuitive.

For example, $e_x = \frac{\partial}{\partial x}$ is a constant vector field on M, so its derivative with respect to any of the coordinates, r or θ, must be identically zero; however, with respect to the basis $\{e_r, e_\theta\}$ we have

$$e_x = \frac{\partial}{\partial x} = \frac{\partial r}{\partial x}\frac{\partial}{\partial r} + \frac{\partial \theta}{\partial x}\frac{\partial}{\partial \theta} = \cos(\theta)\frac{\partial}{\partial r} - \sin(\theta)\frac{1}{r}\frac{\partial}{\partial \theta} = \cos(\theta)e_r - \sin(\theta)e_\theta$$

that is, $e_x = (\cos(\theta), -\sin(\theta))$, so its components are not constant. In fact,

$$\frac{\partial}{\partial \theta}(e_x)^r = \frac{\partial}{\partial \theta}\cos\theta = -\sin\theta \neq 0$$

As this exemplifies, differentiation of the components of a vector field does not necessarily give us the derivative of the field itself; it is necessary to differentiate the basis fields as well.

Given $V = V^r e_r + V^\theta e_\theta$,

$$e_r(V) = \frac{\partial V}{\partial r} = \frac{\partial V^r}{\partial r}e_r + V^r\frac{\partial e_r}{\partial r} + \frac{\partial V^\theta}{\partial r}e_\theta + V^\theta\frac{\partial e_\theta}{\partial r}$$

that is,

$$e_r(V) = \frac{\partial V}{\partial r} = \frac{\partial V^a}{\partial r}e_a + V^a\frac{\partial e_a}{\partial r} \tag{12.6}$$

and similarly for $\frac{1}{r}\frac{\partial V}{\partial \theta} = e_\theta(V)$.

If we identify $e_r(V)$ with $\nabla_{e_r}V$, equation (12.6) tells us that the second term on the right-hand side of the equation represents the connection coefficients, that is,

$$\frac{\partial e_a}{\partial x^b} = \Gamma^c_{ab}e_c \tag{12.7}$$

This would seem to mean that Γ^a_{bc} are not the coefficients of a tensor, since they become zero in one coordinate system (the Cartesian one) and not in the other.

12.6 Theorem (transformation of the connection coefficients)

Under a coordinate change

$$\hat{e}_a = A^b_{\ a}\, e_b$$

we have

$$\hat{\Gamma}^a_{\ bc} = (A^{-1})^a_{\ f}(A)^g_{\ b}(A)^h_{\ c}\Gamma^f_{\ gh} + (A^{-1})^a_{\ f}(A)^h_{\ c}e_h(A)^f_{\ b} \tag{12.8}$$

Proof. We know that

$$\nabla_{\hat{e}_c}\hat{e}_b = \hat{\Gamma}^a_{\ bc}\hat{e}_a = \hat{\Gamma}^a_{\ bc}A^f_{\ a}\, e_f$$

On the other hand,

$$\nabla_{\hat{e}_c}\hat{e}_b = \nabla_{A^h{}_c e_h}(A^g{}_b e_g)$$

$$= A^h{}_c A^g{}_b \nabla_{e_h} e_g + A^h{}_c e_h(A^g{}_b) e_g$$

$$= A^g{}_b A^h{}_c \Gamma^f{}_{gh} e_f + A^h{}_c e_h(A^f{}_b) e_f$$

Therefore,

$$\hat{\Gamma}^a{}_{bc} = (A^{-1})^a{}_f (A)^g{}_b (A)^h{}_c \Gamma^f{}_{gh} + (A^{-1})^a{}_f (A)^h{}_c e_h(A^f{}_b)$$

$\square$

12.7 Example (Γ in polar coordinates)

Let $M = \mathbb{R}^2$ with Cartesian coordinates (x^1, x^2) and basis $\{e_1 = \frac{\partial}{\partial x^1}, e_2 = \frac{\partial}{\partial x^2}\}$. Consider the transformation to coordinates (r, θ) and to the coordinate basis $\{\hat{e}_1 = e_r = \frac{\partial}{\partial r}, \hat{e}_2 = e_\theta = \frac{\partial}{\partial \theta}\}$. We wish to calculate $\hat{\Gamma}^a{}_{bc}$, knowing (from example 12.5) that $\Gamma^a{}_{bc} \equiv 0$. We have $\hat{e}_i = A^j{}_i e_j = A^1{}_i e_1 + A^2{}_i e_2$, $x^1 = r\cos\theta$, and $x^2 = r\sin\theta$. Then

$$\frac{\partial}{\partial r} = \frac{\partial x^1}{\partial r}\frac{\partial}{\partial x^1} + \frac{\partial x^2}{\partial r}\frac{\partial}{\partial x^2} = \cos\theta \frac{\partial}{\partial x^1} + \sin\theta \frac{\partial}{\partial x^2}$$

$$\frac{\partial}{\partial \theta} = \frac{\partial x^1}{\partial \theta}\frac{\partial}{\partial x^1} + \frac{\partial x^2}{\partial \theta}\frac{\partial}{\partial x^2} = -r\sin\theta \frac{\partial}{\partial x^1} + r\cos\theta \frac{\partial}{\partial x^2}$$

Hence,

$$A = \begin{pmatrix} \cos\theta & -r\sin\theta \\ \sin\theta & r\cos\theta \end{pmatrix}$$

and

$$A^{-1} = \begin{pmatrix} \cos\theta & \sin\theta \\ -\dfrac{1}{r}\sin\theta & \dfrac{1}{r}\cos\theta \end{pmatrix}$$

Since $\Gamma^a{}_{bc} \equiv 0$, equation (12.8) implies that

$$\hat{\Gamma}^a{}_{bc} = (A^{-1})^a{}_f A^h{}_c e_h(A^f{}_b) = (A^{-1})^a{}_f \hat{e}_c(A^f{}_b)$$

Therefore,

$$\hat{\Gamma}^\theta{}_{r\theta} = (A^{-1})^\theta{}_f \frac{\partial}{\partial \theta} A^f{}_r = -\frac{1}{r}\sin\theta \frac{\partial}{\partial \theta}\cos\theta + \frac{1}{r}\cos\theta \frac{\partial}{\partial \theta}\sin\theta = \frac{1}{r}$$

$$\hat{\Gamma}^r{}_{\theta\theta} = (A^{-1})^r{}_f \frac{\partial}{\partial \theta} A^f{}_\theta = \cos\theta \frac{\partial}{\partial \theta}[-r\sin\theta] + \sin\theta \frac{\partial}{\partial \theta}[r\cos\theta] = -r$$

etc.

12.8 Lemma (difference of two connections)

Let ∇ and $\bar{\nabla}$ be two connections on M. Then $\nabla - \bar{\nabla}$ is a tensor.

Proof. We define $D(X, Y) \overset{\text{def}}{=} \nabla_X Y - \bar{\nabla}_X Y$. Let $f: M \longrightarrow \mathbb{R} \in C^1$. D inherits the linearity in its first argument from ∇ and $\bar{\nabla}$. We prove its linearity in its second argument directly:

$$
\begin{aligned}
D(X, fY) &= \nabla_X(fY) - \bar{\nabla}_X(fY) \\
&= f\nabla_X Y + (Xf)Y - f\bar{\nabla}_X Y - (Xf)Y \\
&= f(\nabla_X Y - \bar{\nabla}_X Y) \\
&= f\, D(X, Y)
\end{aligned}
$$

$\square$

12.9 Definition (action of ∇ on functions)

Let $f: M \longrightarrow \mathbb{R}$ and $X \in \mathfrak{X}(M)$. Then

$$
\nabla_X f \overset{\text{def}}{=} Xf \tag{12.9}
$$

12.10 Lemma (action of ∇ on 1-forms)

Let $\eta \in \mathfrak{X}^*(M)$. We use the notation $\nabla_a \overset{\text{def}}{=} \nabla_{e_a}$. Leibnitz's rule allows us to extend ∇_a to 1-forms, and

$$
(\nabla_a \eta)_b = e_a(\eta_b) - \Gamma^c{}_{ba}\eta_c \tag{12.10}
$$

Proof. Let $\{e_a\}$ be a basis of $\mathfrak{X}(M)$ and $\{\omega^a\}$ be its dual basis on $\mathfrak{X}^*(M)$. Then $\eta = \eta_c \omega^c$ and $\nabla_a [\eta(e_b)] = e_a [\eta(e_b)]$. Using Leibnitz's rule, we have

$$
\nabla_a [\eta(e_b)] = (\nabla_a \eta)e_b + \eta\nabla_a e_b \tag{12.11}
$$

Therefore,

$$
\begin{aligned}
(\nabla_a \eta)_b &= (\nabla_a \eta)e_b \\
&= e_a [\eta(e_b)] - \eta\nabla_a e_b \\
&= e_a(\eta_b) - \Gamma^c{}_{ba}\eta_c
\end{aligned}
$$

$\square$

Note that, if $\{e_a\}$ is a coordinate basis, then

$$
\eta_{b;a} \overset{\text{def}}{=} (\nabla_a \eta)_b = \eta_{b,\,a} - \Gamma^c{}_{ba}\eta_c
$$

12.11 Lemma (action of ∇ on tensors)

Leibnitz's rule allows us to extend ∇_a to rank-$\binom{1}{1}$ tensors. Let T be a rank-$\binom{1}{1}$ tensor on M. Then

$$(\nabla_a T)_b^{\ c} = e_a(T_b^{\ c}) + \Gamma^c_{fa} T_b^{\ f} - \Gamma^d_{ba} T_d^{\ c} \tag{12.12}$$

Proof. Let $\{e_a\}$ be a basis of $\mathfrak{X}(M)$ and $\{\omega^a\}$ be its dual basis in $\mathfrak{X}^*(M)$. Then

$$\nabla_a [T(e_b, \omega^c)] = e_a [T(e_b, \omega^c)]$$

On the other hand, using Leibnitz's rule, we have

$$\nabla_a [T(e_b, \omega^c)] = (\nabla_a T)(e_b, \omega^c) + T(\nabla_a e_b, \omega^c) + T(e_b, \nabla_a \omega^c)$$

so

$$(\nabla_a T)(e_b, \omega^c) = e_a [T(e_b, \omega^c)] - T(\nabla_a e_b, \omega^c) - T(e_b, \nabla_a \omega^c)$$

that is,

$$(\nabla_a T)_b^{\ c} = e_a T_b^{\ c} - T(\Gamma^d_{ba} e_d, \omega^c) + T(e_b, \Gamma^c_{fa} \omega^f)$$
$$= e_a T_b^{\ c} - \Gamma^d_{ba} T_d^{\ c} + \Gamma^c_{fa} T_b^{\ f}$$

$\square$

Lemma 12.11 is trivially generalised to tensors of arbitrary rank.

12.12 Example ('orthogonality' condition)

Let ∇ be a 'symmetric connection' on M, that is, $\Gamma^c_{ba} = \Gamma^c_{ab}$. Let $X \in \mathfrak{X}(M)$ be a vector field which is orthogonal to a family of hypersurfaces; we still do not know what 'orthogonality' means, since we have not yet defined an inner product on M; let us therefore take 'orthogonality' to mean that, if the family of hypersurfaces is given by $f = $ const., then

$$X \propto \nabla f$$

that is, there exists $h: M \longrightarrow \mathbb{R}$ such that $X^i = h \, \delta^{ij} f_{,j} = h f^{,i}$.

Then

$$X^i_{\ ;k} = h_{,k} f^{,i} + h f^{,i}_{\ ;k}$$

and

$$X^i_{\ ;k} X^l = h_{,k} f^{,i} X^l + h f^{,i}_{\ ;k} X^l$$

Substituting $X^l = h f^{,l}$ on the right-hand side and antisymmetrising in all three indices i, k, l,

$$X^{[i}_{\ ;k}X^{l]} = h_{,[k}f^{,i}hf^{,l]} + h^2 f^{,[i}_{\ ;k}f^{,l]}$$

The first term on the right-hand side is zero because of the antisymmetrisation, and the second term is zero because ∇ is symmetric and hence $f^{,i}_{\ ;k} = f_{,k}^{\ ;i}$. Therefore, the fact that X is orthogonal to the hypersurfaces implies that

$$X^{[i}_{\ ;k}X^{l]} = 0 \tag{12.13}$$

with antisymmetrisation in all indices. The converse of this result is called 'Frobenius' theorem'.

IOP Publishing

Differential Topology and Geometry with Applications to Physics

Eduardo Nahmad-Achar

Chapter 13

Geodesics

The concept of a 'geodesic' is a generalisation of the straight line in a flat space, that is, the 'minimum-length' curve between two points. Once we generalise the 'minimum-length curve' concept to manifolds, it becomes valid only locally; however, one property of straight lines in flat spaces, that of parallel-transporting their tangent vectors, is generalised in a direct way.

The study of a manifold's geodesics is important because they are the paths followed by free particles. Since general relativity replaces the 'force of gravity' with the curvature of space, any object moving in the presence of gravitational fields follows a geodesic path.

13.1 Definition (geodesic)

Let $X \in \mathfrak{X}(M)$. The integral curves of X are called 'geodesics' if and only if

$$\nabla_X X = 0 \qquad (13.1)$$

13.2 Theorem (uniqueness of the geodesic given a point and a tangent vector)

Given $p \in M$ and $X_p \in T_p(M)$, there is a unique geodesic which passes through p and has X_p as tangent vector at p.

Proof. Let us take a local coordinate system $\{x^i\}$ in a neighbourhood of p. The vector field X tangent to the curve would be given by

$$\frac{dx^i(t)}{dt} = X^i[x(t)] \qquad (13.2)$$

doi:10.1088/2053-2563/aadf65ch13

where t is the parameter along the curve, and the condition that this curve be a geodesic is

$$0 = (\nabla_X X)^i = X^i_{;j} X^j = X^i_{,j} X^j + \Gamma^i_{jk} X^j X^k$$

whereby

$$\Gamma^i_{jk} X^j X^k = -X^i_{,j} X^j = -\frac{dX^i}{dt} \tag{13.3}$$

that is, using equation (13.2),

$$\frac{d^2 x^i(t)}{dt^2} = -\Gamma^i_{jk} \frac{dx^j}{dt} \frac{dx^k}{dt} \tag{13.4}$$

which we must solve subject to the initial conditions

$$x^i(0) = x^i_p, \qquad \frac{dx^i}{dt}(0) = X^i_p$$

The theorem of existence and uniqueness of solutions to ordinary differential equations guarantees that, for a sufficiently small interval of t, the geodesic we seek exists and is unique. $\qquad\square$

13.3 Theorem (uniqueness of the geodesic given two points)

Given a sufficiently small neighbourhood $U \subset M$ and two points $p, q \in U$ such that $p \neq q$, there is a unique geodesic γ which passes through both p and q and satisfies $\gamma(0) = p$, $\gamma(1) = q$.

Proof. As before; in this case, we must solve equation (13.4) with boundary conditions

$$x^i(0) = x^i_p$$
$$x^i(1) = x^i_q$$

$\qquad\square$

13.4 Exercise (affine parameter)

As seen in the proof of theorem 13.2, the equation of a geodesic in a local coordinate system is

$$\ddot{x}^i + \Gamma^i_{jk} \dot{x}^j \dot{x}^k = 0 \tag{13.5}$$

where $\dot{x}^i = \frac{dx^i}{dt}$. Suppose we make a parameter change $t \longmapsto s(t)$. What sort of transformations of the parameter preserve equation (13.5)? These parameters are called 'affine'.

Solution. Let us transform $t \longmapsto s(t)$, and let $x'^i = \frac{dx^i}{ds}$ (i.e. $\dot{x} \equiv \frac{d}{dt}x$, $x' \equiv \frac{d}{ds}x$). Then

$$x'^i = \frac{dx^i}{ds} = \frac{dx^i}{dt}\frac{dt}{ds}$$

that is,

$$\dot{x}^i = \dot{s}x'^i$$
$$\ddot{x}^i = \dot{s}^2 x''^i + \ddot{s}x'^i$$

Equation (13.5) then becomes

$$\dot{s}^2(x''^i + \Gamma^i_{jk}x'^j x'^k) = -\ddot{s}x'^i \tag{13.6}$$

In other words, equation (13.5) preserves its form if and only if $\ddot{s} = 0$, which is to say that $s = at + b$ for some constant numbers a and b.

13.5 Definition (exponential map and normal coordinates)

(i) Let $p \in M$, and for each $X_p \in T_p(M)$ let γ_X be the (only) geodesic such that $\gamma_X(0) = p$ and $\dot{\gamma}_X(0) = X_p$. Then

$$\exp : T_p(M) \longrightarrow M$$
$$X_p \longmapsto \gamma_X(1)$$

is called the 'exponential map' on M.

(ii) Theorem 13.3 guarantees that exp has an inverse map in a small neighbourhood of p : if $\{e_a\}$ is a basis of $T_p(M)$, there is a neighbourhood U of p such that, for all $q \in U$, $\exists X_p \in T_p(M)$ such that

$$q = \exp(X_p) = \exp(X_p^a e_a)$$

$\{X_p^a\}_{a \in \{1, \ldots, n\}}$ are called 'normal coordinates' of q with respect to $\{e_a\}$.

An interesting case occurs when M is a Lie group; in addition, this justifies the name *exponential map*. The most important class of Lie groups in physics is that of the matrix groups, and the most general of these groups is $GL(n, \mathbb{F})$, the invertible $n \times n$ matrices over a field $\mathbb{F}$ (which is usually $\mathbb{R}$ or $\mathbb{C}$):

$$GL(n, \mathbb{F}) = \{A \text{ such that } A \text{ is an } n \times n \text{ matrix over } \mathbb{F} \text{ and } \det(A) \neq 0\}.$$

Its associated Lie algebra is (*vide infra*)

$$gl(n, \mathbb{F}) = \{n \times n \text{ matrices over } \mathbb{F}\}$$

with $[A, B] = AB - BA$.

In this case, the exponential map is a map from the Lie algebra to its associated Lie group. Recall that, for $A \in gl(n, \mathbb{F})$,

$$e^A = \exp(A) = \sum_{k=0}^{\infty} \frac{A^k}{k!}$$

Then

$$\gamma_A : \mathbb{R} \longrightarrow GL(n, \mathbb{F})$$
$$t \longmapsto e^{tA}$$

is a soft curve on $GL(n, \mathbb{F}) \subset \mathbb{F}^{n^2}$ which respects the group structure:

$$\gamma_A(t_1 + t_2) = \gamma_A(t_1)\gamma_A(t_2)$$

that is, it is a group homomorphism. The corresponding algebra homomorphism (see exercise 9.8) is the differential map

$$\gamma_{A*}(t) = \frac{d}{dt} e^{tA}$$

Evaluating this map at the origin, we obtain $\frac{d}{dt} e^{tA}|_{t=0} = A$, which shows that the space tangent to the identity ($t = 0$) is indeed $gl(n, \mathbb{F})$.

The Lie-group homomorphism γ_A is unique because $\mathbb{R}$ is connected, and the result is important because it allows us to obtain in a simple way a representation of a Lie group once a representation of its associated Lie algebra is available.

13.6 Theorem (symmetric part of the connection)

For all $p \in M$ there is a coordinate system $\{x^a\}$ such that

$$\Gamma^i_{(jk)} = 0$$

at p.

Proof. We use the notation from definition 13.5. Let $p \in M$, $U \subset M$ be a neighbourhood of p, $\{x^a\}$ the normal coordinates in U with respect to p, $q \in U$ be such that $q \neq p$, and γ_X be the geodesic passing through p and q and satisfying $\gamma_X(0) = p$ and $\dot{\gamma}_X(0) = X_p \in T_p(M)$.

By construction, the normal coordinates of any point on the geodesic must be proportional to X_p^a, that is, for all $r \in \gamma_X$ $\exists s \in \mathbb{R}$ such that

$$x^a(r) = s X_p^a$$

Let t be an affine parameter of γ_X such that $\dot{s}(0) = \frac{ds}{dt}(0) \neq 0$. Then

$$x'^j \overset{\text{def}}{=} \frac{dx^j}{ds} = X^j$$
$$x''^j = \frac{d^2 x^j}{ds^2} = 0$$

and, from equation (13.6), the equation for γ_X is

$$\dot{s}^2 \Gamma^i_{jk} X^j X^k = -\ddot{s} X^i$$

Therefore, at p we have

$$\Gamma^i_{jk}(p) X^j_p X^k_p = -\left(\frac{\ddot{s}}{\dot{s}^2}\right)(0) X^i_p$$

Repeating the same construction for $-X_p$, we necessarily have

$$\Gamma^i_{(jk)}(p) = 0$$

$\square$

13.7 Example (connection in polar coordinates)

The general equation for a geodesic is

$$\frac{d^2 x^i}{ds^2} + \Gamma^i_{jk} \frac{dx^j}{ds} \frac{dx^k}{ds} = 0 \tag{13.7}$$

In $\mathbb{E}^2$ with Cartesian coordinates (x^1, x^2) we have $\Gamma^i_{jk} \equiv 0$. Thus,

$$\frac{d^2 x^1}{ds^2} = \frac{d^2 x^2}{ds^2} = 0$$

which represents straight lines with constant velocity.

Transforming the geodesic equation $\frac{d^2 x^i}{ds^2} = 0$ to polar coordinates

$$x^1 = r \cos \theta$$
$$x^2 = r \sin \theta$$

we obtain

$$\frac{dx^1}{ds} = \frac{\partial x^1}{\partial r} \frac{dr}{ds} + \frac{\partial x^1}{\partial \theta} \frac{d\theta}{ds} = \cos \theta \frac{dr}{ds} - r \sin \theta \frac{d\theta}{ds}$$

$$\frac{dx^2}{ds} = \frac{\partial x^2}{\partial r} \frac{dr}{ds} + \frac{\partial x^2}{\partial \theta} \frac{d\theta}{ds} = \sin \theta \frac{dr}{ds} + r \cos \theta \frac{d\theta}{ds}$$

and

$$\frac{d^2 x^1}{ds^2} = \frac{\partial}{\partial r}\left(\frac{dx^1}{ds}\right)\frac{dr}{ds} + \frac{\partial}{\partial \theta}\left(\frac{dx^1}{ds}\right)\frac{d\theta}{ds}$$

$$= \cos \theta \frac{d^2 r}{ds^2} - 2 \sin \theta \frac{dr}{ds} \frac{d\theta}{ds} - r \cos \theta \frac{d\theta}{ds} \frac{d\theta}{ds} - r \sin \theta \frac{d^2\theta}{ds^2} \tag{13.8}$$

$$\frac{d^2x^2}{ds^2} = \frac{\partial}{\partial r}\left(\frac{dx^2}{ds}\right)\frac{dr}{ds} + \frac{\partial}{\partial \theta}\left(\frac{dx^2}{ds}\right)\frac{d\theta}{ds}$$

$$= \sin\theta\frac{d^2r}{ds^2} + 2\cos\theta\frac{dr}{ds}\frac{d\theta}{ds} - r\sin\theta\frac{d\theta}{ds}\frac{d\theta}{ds} + r\cos\theta\frac{d^2\theta}{ds^2} \tag{13.9}$$

Both equations (13.8) and (13.9) must be equal to zero. Multiplying equation (13.8) by $\cos\theta$ and equation (13.9) by $\sin\theta$ and adding them,

$$0 = \frac{d^2}{ds^2} - r\frac{d\theta}{ds}\frac{d\theta}{ds} \tag{13.10}$$

Multiplying equation (13.8) by $\sin\theta$ and equation (13.9) by $\cos\theta$ and subtracting one from the other,

$$0 = -2\frac{dr}{ds}\frac{d\theta}{ds} - r\frac{d^2\theta}{ds^2}$$

that is,

$$0 = \frac{d^2\theta}{ds^2} + \frac{2}{r}\frac{dr}{ds}\frac{d\theta}{ds} \tag{13.11}$$

Therefore, reading directly from equations (13.7), (13.10), and (13.11), the connection coefficients are

$$\Gamma^r{}_{\theta\theta} = -r$$
$$\Gamma^\theta{}_{r\theta} = \Gamma^\theta{}_{\theta r} = \frac{1}{r} \tag{13.12}$$
$$\Gamma^i{}_{jk} = 0 \quad \text{in all other cases}$$

13.8 Example (Newtonian potential)

At the beginning of this chapter we mentioned that in general relativity, gravitational phenomena are considered a consequence of the curvature of space and a particle subject to no external forces moves along a geodesic.

In the Newtonian limit, let us denote by

$$x^\alpha = (t, x^1, x^2, x^3) = (x^0, x^i)_{i=1, 2, 3}$$

the coordinates of a moving particle in a neighbourhood U of a point $p \in M$. The particle velocity is

$$v^\alpha = \frac{dx^\alpha}{dt} = (1, \vec{v})$$

where $\vec{v}$ is the usual three-dimensional velocity.

If the particle moves in a gravitational field with Newtonian potential φ, its equations of motion are

$$\ddot{x}^\alpha = (0, -\nabla\varphi) = (0, -\varphi^{,1}, -\varphi^{,2}, -\varphi^{,3}) = -\varphi^{,\alpha}$$

where we have set $\varphi^{,0} = 0$ and, since $\dot{x}^0 = 1$, we have

$$\ddot{x}^\alpha + \varphi^{,\alpha}\dot{x}^0\dot{x}^0 = 0 \tag{13.13}$$

$$\ddot{x}^0 = 0 \tag{13.14}$$

which is the equation of a geodesic. The coefficients of the 'Newtonian connection' are

$$\begin{aligned}
\Gamma^k_{\ 00} &= \varphi_{,k} \\
\Gamma^i_{\ jk} &= 0 \quad \text{in all other cases}
\end{aligned} \tag{13.15}$$

IOP Publishing

Differential Topology and Geometry with Applications to Physics

Eduardo Nahmad-Achar

Chapter 14

Torsion and curvature

We have built all the necessary mathematical tools (and a few more) to define and study the concepts of torsion and curvature of a manifold. While in the case of one- and two-dimensional manifolds a scalar suffices to describe either of these quantities at each point, for manifolds of larger dimension the geometry around a point is considerably richer and we require tensors to describe them.

So far, we have constructed two vector fields, $[X, Y]$ and $\nabla_X Y - \nabla_Y X$, both of which are antisymmetric in X and Y.

14.1 Definition (symmetric connection)

A linear connection ∇ on M is called 'symmetric' if and only if

$$[X, Y] = \nabla_X Y - \nabla_Y X$$

for all $X, Y \in \mathfrak{X}(M)$.

14.2 Lemma (justification for definition 14.1)

A linear connection ∇ is symmetric if and only if

$$\Gamma^k{}_{lm} = \Gamma^k{}_{ml}$$

Proof. We write

$$(\nabla_X Y - \nabla_Y X)^k = Y^k{}_{,a}X^a + \Gamma^k{}_{lm}Y^l X^m - X^k{}_{,a}Y^a - \Gamma^k{}_{lm}X^l Y^m$$

and

$$[X, Y]^k = (XY)^k - (YX)^k = Y^k{}_{,a}X^a - X^k{}_{,a}Y^a$$

whereby

$$(\nabla_X Y - \nabla_Y X)^k = [X, Y]^k \Leftrightarrow \Gamma^k{}_{lm}Y^l X^m - \Gamma^k{}_{lm}X^l Y^m = 0$$

But

$$\Gamma^k{}_{lm} Y^l X^m - \Gamma^k{}_{lm} X^l Y^m = (\Gamma^k{}_{lm} - \Gamma^k{}_{ml}) Y^l X^m$$

that is,

$$(\nabla_X Y - \nabla_Y X)^k = [X, Y]^k \Leftrightarrow \Gamma^k{}_{lm} = \Gamma^k{}_{ml}$$

$\square$

14.3 Definition (torsion coefficients)

Let $\{e_a\}$ be a basis of $\mathfrak{X}(M)$, and let $\dim(M) = n$. $\nabla_i e_j$, $\nabla_j e_i$ and $[e_i, e_j]$ are vector fields, so for all $i, j, k \in \{1, \dots, n\}$ we can define

$$\nabla_i e_j - \nabla_j e_i - [e_i, e_j] \overset{\text{def}}{=} T^k{}_{ij} e_k \tag{14.1}$$

The $T^k{}_{ij}$ are called 'torsion coefficients'.

14.4 Theorem (torsion tensor)

The torsion coefficients $T^k{}_{ij}$ are the components of a tensor.

Proof. We define

$$T: \mathfrak{X}(M) \times \mathfrak{X}(M) \longrightarrow \mathfrak{X}(M)$$
$$(X, Y) \longmapsto \nabla_X Y - \nabla_Y X - [X, Y]$$

Let $f: M \longrightarrow \mathbb{R} \in C^1$. Then for all $X, Y, Z \in \mathfrak{X}(M)$ we have

$$T(X + Z, Y) = T(X, Y) + T(Z, Y)$$

and

$$\begin{aligned}
T(fX, Y) &= \nabla_{fX} Y - \nabla_Y(fX) - [fX, Y] \\
&= f\nabla_X Y - f\nabla_Y X - Y(f)X - f[X, Y] + Y(f)X \\
&= fT(X, Y)
\end{aligned}$$

and, since $T(X, Y) = -T(Y, X)$, T is also linear in its second argument. Therefore,

$$\hat{T}: \mathfrak{X}(M) \times \mathfrak{X}(M) \times \mathfrak{X}^*(M) \longrightarrow \mathbb{R}$$
$$(X, Y, \omega) \longmapsto \langle \omega, T(X, Y) \rangle$$

is a tensor and its components are $T^k{}_{ij}$ by construction. $\square$

It is customary to also call T, not only $\hat{T}$, the 'torsion tensor'.

14.5 Lemma (antisymmetric part of the connection)

Let $\{e_a\}$ be a basis of $\mathfrak{X}(M)$, and let γ^c_{ab} be structure functions, that is, $[e_a, e_b] = \gamma^c_{ab}e_c$. Then

$$T^c_{ab} = -2\Gamma^c_{[ab]} - \gamma^c_{ab} \tag{14.2}$$

Proof. We have

$$\begin{aligned}
T(e_a, e_b) &= \nabla_a e_b - \nabla_b e_a - [e_a, e_b] \\
&= (\Gamma^c_{ba} - \Gamma^c_{ab} - \gamma^c_{ab})e_c \\
&= (-2\Gamma^c_{[ab]} - \gamma^c_{ab})e_c
\end{aligned}$$

$\square$

Theorem 13.6 tells us that geodesics fix the symmetric part of the connection. Lemma 14.5 tells us that torsion represents the part of the connection which remains undetermined by the geodesics.

General relativity assumes that the connection is introduced precisely to describe certain curves as geodesics, and thus it assumes that there is no torsion. There exist, however, other theories of gravitation which give a physical meaning to torsion. The best-known of these is the Einstein–Cartan theory.

14.6 Theorem (curvature tensor)

Let

$$R: \mathfrak{X}(M) \times \mathfrak{X}(M) \times \mathfrak{X}(M) \longrightarrow \mathfrak{X}(M)$$
$$(X, Y, Z) \longmapsto \nabla_X \nabla_Y Z - \nabla_Y \nabla_X Z - \nabla_{[X, Y]}Z$$

Then

$$\hat{R}: \mathfrak{X}(M) \times \mathfrak{X}(M) \times \mathfrak{X}(M) \times \mathfrak{X}^*(M) \longrightarrow \mathbb{R}$$
$$(X, Y, Z, \omega) \longmapsto \langle \omega, R(X, Y, Z) \rangle$$

is a tensor, the 'curvature tensor', also called the 'Riemann tensor'. It is usual to call R the 'curvature tensor' or the 'Riemann tensor' as well.

Proof. By definition, $R(X, Y, Z) = -R(Y, X, Z)$, so it suffices to prove that R is linear in its first and third arguments. Linearity under addition and under multiplication by a scalar are trivial.

Let $f: M \longrightarrow \mathbb{R} \in C^2$. Then

$$\begin{aligned}
R(fX, Y, Z) &= \nabla_{fX} \nabla_Y Z - \nabla_Y \nabla_{fX} Z - \nabla_{[fX, Y]}Z \\
&= f\nabla_X \nabla_Y Z - f\nabla_Y \nabla_X Z - Y(f)\nabla_X Z - \nabla_{(f[X, Y] - Y(f)X)}Z \\
&= fR(X, Y, Z)
\end{aligned}$$

and

$$R(X, Y, fZ) = \nabla_X \nabla_Y (fZ) - \nabla_Y \nabla_X (fZ) - \nabla_{[X, Y]}(fZ)$$
$$= \nabla_X (f \nabla_Y Z + Y(f)Z) - \nabla_Y (f \nabla_X Z + X(f)Z)$$
$$- f \nabla_{[X, Y]} Z - ([X, Y]f)Z$$
$$= fR(X, Y, Z) + X(f)\nabla_Y Z + XY(f)Z + Y(f)\nabla_X Z$$
$$- Y(f)\nabla_X Z - YX(f)Z - X(f)\nabla_Y Z - ([X, Y]f)Z$$
$$= fR(X, Y, Z)$$

Therefore R is linear in all of its arguments and, since $\hat{R}$ is by construction linear in its last argument as well, $\hat{R}$ is a tensor. $\qquad\square$

14.7 Lemma (calculation of the curvature tensor)

Let $\{e_a\}$ be a basis of $\mathfrak{X}(M)$ and define

$$R(e_c, e_d, e_b) \stackrel{\text{def}}{=} R^a{}_{bcd} e_a \qquad (14.3)$$

Then

$$R^a{}_{bcd} = e_c(\Gamma^a{}_{bd}) - e_d(\Gamma^a{}_{bc}) + \Gamma^f{}_{bd}\Gamma^a{}_{fd} - \gamma^f{}_{cd}\Gamma^a{}_{bf} \qquad (14.4)$$

Proof. $R^a{}_{bcd}$ are, by definition, the components of the Riemann tensor. From the definition, we have

$$R(e_c, e_d, e_b) = \nabla_c \nabla_d e_b - \nabla_d \nabla_c e_b - \nabla_{[e_c, e_d]} e_b$$
$$= \nabla_c(\Gamma^f{}_{bd} e_f) - \nabla_d(\Gamma^f{}_{bc} e_f) - \gamma^f{}_{cf} \nabla_f e_b$$
$$= e_c(\Gamma^a{}_{bd})e_a + \Gamma^f{}_{bd}\Gamma^a{}_{fc}e_a - e_d(\Gamma^a{}_{bc})e_a + \Gamma^f{}_{bc}\Gamma^a{}_{fd}e_a - \gamma^f{}_{cd}\Gamma^a{}_{bf}e_a$$

Therefore, using equation (14.3),

$$R^a{}_{bcd} = e_c(\Gamma^a{}_{bd}) - e_d(\Gamma^a{}_{bc}) + \Gamma^f{}_{bd}\Gamma^a{}_{fd} - \gamma^f{}_{cd}\Gamma^a{}_{bf} \qquad (14.4)$$

$\qquad\square$

14.8 Geometric interpretations (see chapter 2)

In this section we shall repeat some concepts which were already introduced in chapter 2; this is to make the current chapter self-contained and for the benefit of readers who have skipped chapter 2.

Let $\alpha(t)$ be a curve in $\mathbb{R}^3$. The 'arc length' of α from $t = a$ to $=b$ is defined as

$$\int_a^b \|\alpha'(t)\| \, dt$$

where $\alpha = (\alpha_1, \alpha_2, \alpha_3)$ and $\alpha'(t) = (\frac{d\alpha_1}{dt}(t), \frac{d\alpha_2}{dt}(t), \frac{d\alpha_3}{dt}(t))$ in Euclidean coordinates.

When we are only interested in the curve's trajectory and not in the speed with which it has travelled, it is convenient to parametrise the curve in such a way that its speed is unity, that is, $||\alpha'|| = 1$. This is achieved considering the arc-length function

$$s(t) = \int_a^t ||\alpha'(u)|| \, du$$

where a is any fixed point in the domain of α. It is clear that $\frac{ds}{dt} = ||\alpha'||$. In addition, if α is 'regular' (that is, $\alpha' \neq 0$ for all t), we may write, without loss of generality, $\frac{ds}{dt} > 0$, whereby $s(t)$ is invertible: $t = t(s)$. Let $\beta(s) = \alpha[t(s)]$. Then $\beta' = \frac{dt}{ds}(s)\alpha'[t(s)]$, that is,

$$||\beta'(s)|| = \frac{dt}{ds}(s)||\alpha'(t)|| = \frac{dt}{ds}(s)\frac{ds}{dt}[t(s)] = 1 \tag{14.5}$$

Therefore, $\beta(s)$ is the 'parametrisation by arc length' of $\alpha(t)$.

Now let $\alpha(t)$ be a curve in $\mathbb{R}^3$ parametrised by arc length. We define

$$T = \alpha' \text{ (vector field tangent to the curve)}$$

$$N = \frac{T'}{||T'||} \text{ (vector field normal to the curve)}$$

$$B = T \times N \text{ (vector field binormal to the curve)}$$

(i) Since $||T|| = 1$, differentiating $T \cdot T = 1$ we obtain $2T' \cdot T = 0$ and thus $T' \perp T$.

(ii) $\kappa(t) = ||T'(t)|| = ||\alpha''(t)||$ is called the 'curvature function' of α.

(iii) $||N|| = 1$ by construction.

(iv) Differentiating $B \cdot T = 0$, we have $0 = B' \cdot T + B \cdot T' = B' \cdot T + B \cdot \kappa N = B' \cdot T$. Since $B \cdot B = 1$, we have $B' \cdot B = 0$. Therefore, there is a function $\tau(t)$ such that $B' = -\tau N$.
τ is called the 'torsion function' of α.

Theorem (Frenet's formulae)

If α is a curve parametrised by arc length, then

$$T' = \kappa N$$
$$N' = -\kappa T + \tau B$$
$$B' = -\tau N$$

This theorem was proven in chapter 2. In that chapter, we also proved that a curve is flat (that is, it is completely contained in a plane) if and only if $\tau = 0$; in other words, a curve's torsion indicates how far from being flat that curve is. Similarly, the curve's curvature κ indicates how far the curve is from being a straight line (see lemma 2.16). Figure 14.1 shows the natural reference frame field (T, N, B) for $\alpha(t)$.

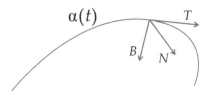

Figure 14.1. Tangent, normal, and binormal to a curve $\alpha(t) \in \mathbb{R}^3$ parametrised by arc length. They constitute a natural reference frame field for the curve.

14.8.1 Geometric interpretation of torsion for manifolds of dimension higher than 2

We can interpret the torsion tensor on a manifold of arbitrary dimension as follows.

Recall that the geometric interpretation of the commutator of two vector fields is that it 'closes the quadrilateral' determined by the vector fields (see chapter 7). Let p be a point on a manifold M, and let X and Y be two vector fields such that X_p and Y_p are nonparallel. Figure 14.2 shows geometrically the effect of the commutator, where we assume that p, p_1, and p_2 are infinitesimally close.

Suppose there is no torsion, that is, $T^i_{jk} \equiv 0$. Let us parallel-transport Y_p along X to p_1 and X_p along Y to p_2. Let $X_\parallel$ and $Y_\parallel$ be, respectively, the vectors so obtained. Figure 14.2 shows these vectors, as well as the differences $Y_{p_1} - Y_\parallel = \nabla_X Y$ and $X_{p_2} - X_\parallel = \nabla_Y X$.

From the figure, we have

$$\nabla_X Y - \nabla_Y X = [X, Y]$$

that is,

$$\nabla_X Y - \nabla_Y X - [X, Y] = 0 \tag{14.6}$$

which is the condition that the torsion be zero.

Now, if there is torsion, this will affect the parallel transport of the fields and we will obtain something similar to that shown in figure 14.3.

If V is the vector which closes the quadrilateral formed by X_p, Y_p, $X_\parallel$, and $Y_\parallel$, then

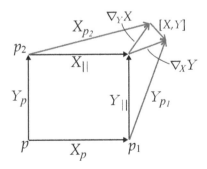

Figure 14.2. Effect of the commutator of two vector fields X and Y when the torsion is zero.

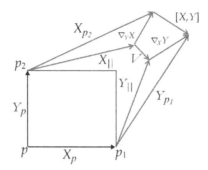

Figure 14.3. Effect of the commutator of two vector fields X and Y when the torsion is non-zero.

$$V = \nabla_X Y - \nabla_Y X - [X, Y] = T(X, Y)$$

that is, V is precisely the vector obtained by applying the torsion tensor to the fields X and Y.

14.8.2 Geometric interpretation of curvature for two-dimensional surfaces

Let M be a two-dimensional surface in $\mathbb{R}^3$ and P be a plane intersecting M. Let α be the curve formed by the intersection between P and M, and suppose it is parametrised by arc length. Finally, let $U(p)$ be the unit vector normal to $\alpha(t)$ at a point p, which is contained in P. Then

$$k(v) \stackrel{\text{def}}{=} - v \cdot \nabla_v U$$

is called the 'normal curvature of M in the direction v' at p (see figure 14.4).

If κ is the curvature of $\alpha(t)$ defined above, and θ is the angle between U and N, the vector field normal to α, then it can be shown that

$$k = \kappa \cos \theta$$

The maximum and minimum values of k at a point $p \in M$ are called the 'principal curvatures of M' at p, and their corresponding directions are called 'principal directions of M' at p.

Let $k_1(p)$ and $k_2(p)$ be the maximum and minimum values of k at p, respectively. Then

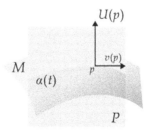

Figure 14.4. Geometric interpretation of the curvature for surfaces in $\mathbb{R}^3$ as the variation of the normal vector field U.

$$K(p) = \det \begin{pmatrix} k_1(p) & 0 \\ 0 & k_2(p) \end{pmatrix}$$

is called the 'Gaussian curvature of M' at p and

$$H(p) = \frac{1}{2}\mathrm{tr}\begin{pmatrix} k_1(p) & 0 \\ 0 & k_2(p) \end{pmatrix}$$

is called the 'mean curvature of M' at p.

The Gaussian curvature is the 'intrinsic curvature' of the surface M and is independent of U. The Riemann tensor is the generalisation of the concept of Gaussian curvature to manifolds of arbitrary dimension. Its geometric interpretation for arbitrary dimensions will now follow.

14.9 Geometric interpretation of curvature

Let $\alpha_s(t)$ be a one-parameter family of curves which pass through a point $p = \alpha_s(t = 0) \in M$.

As s varies, the points $t = \mathrm{const.}$ define a curve $\alpha_t(s) \in M$. For $t = 0$, $\alpha_t(s)$ is degenerated to the point p (see figure 14.5).

Let V_T and V_S be the vector fields tangent to the curves $\alpha_s(t)$ and $\alpha_t(s)$, respectively.

By construction, and from the geometric interpretation of the commutator, we have $[V_T, V_S] = 0$. Let $X_p \in T_p(M)$, and let us define a vector field X as an extension of X_p as follows: given a point (s, t), we parallel-transport X_p along $\alpha_s(t)$, that is,

$$\nabla_{V_T} X = 0$$

$$X(p) = X_p$$

Let $C_{ds,\, dt}$ be the closed curve determined by $pqrp$ in figure 14.5, and let $X_p(ds, dt)$ be the vector at p obtained by parallel-transporting X_p along $C_{ds,\, dt}$ from p to p.

Let us now calculate $X_p(ds, dt) - X_p$.

By construction, $\overline{pq}$ and $\overline{rp}$ do not contribute, as $\nabla_{V_T} X = 0$. Therefore, to first order,

$$X_p(ds, dt) - X_p = X_r - X_q = (\nabla_{V_S} X)ds$$

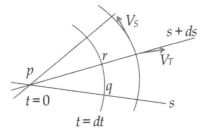

Figure 14.5. One-parameter family of curves which pass through point $p = \alpha_s(t = 0) \in M$. The Riemann tensor at p measures the change in a vector field X which is parallel-transported along the closed curve $pqrp$.

But $V_S = 0$ at p, so $\nabla_{V_S} X = 0$ at p.

Expanding $\nabla_{V_S} X$ to first order in t and setting $\nabla_{V_S} X = 0$ at p,

$$\nabla_{V_S} X = (\nabla_{V_T} \nabla_{V_S} X) dt$$

But $\nabla_{V_T} \nabla_{V_S} X = R(V_T, V_S, X)$, since $[V_T, V_S] = 0$ and $\nabla_{V_T} X = 0$. Hence,

$$X_p(ds, dt) - X_p = R(V_T, V_S, X)_p ds\, dt$$

to first order in s and t. To put it another way,

$$R(V_T, V_S, X)_p = \lim_{\substack{ds \to 0 \\ dt \to 0}} \frac{1}{ds} \frac{1}{dt} \left[X_p(ds, dt) - X_p \right]$$

This means that $R(V_T, V_S, X)$ measures the change in X produced by parallel transport along a closed curve determined by V_T and V_S.

The foregoing leads to the following lemma.

14.10 Lemma (Ricci's identity)

We use the notation $X^i_{\;;jk} = X^i_{\;;j;k}$ for $X \in \mathfrak{X}(M)$. Then

$$X^i_{\;;jk} - X^i_{\;;kj} = R^i_{\;mkj} X^m \tag{14.7}$$

where X^i denotes the ith component of X in a coordinate basis $\left\{ \frac{\partial}{\partial x^i} \right\}$.

Proof. Let $e_i = \frac{\partial}{\partial x^i}$. Then

$$
\begin{aligned}
R^i_{\;mkj} X^m e_i \quad &= \quad R(e_k, e_j, X) \\
&= \quad \nabla_k \nabla_j X - \nabla_j \nabla_k X - \nabla_{[e_k, e_j]} X \\
&\overset{\text{coord.basis}}{=} \quad X_{;jk} - X_{;kj}
\end{aligned}
$$

$\square$

For a different interpretation of the Riemann tensor let us consider the following:

14.11 Definition (field connecting a congruence)

Let X be a vector field on M. Its integral curves define a congruence of curves on M. $Z \in \mathfrak{X}(M)$ is called a 'field connecting the congruence' (see figure 14.6) if and only if

$$[X, Z] = 0$$

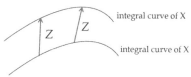

Figure 14.6. Vector field Z connects the congruence of integral curves of vector field X iff $[X, Z] = 0$.

14.12 Theorem (another geometric interpretation of curvature)

Let the torsion vanish, $T = 0$, and let X be a geodesic vector field in M. Let Z be a field connecting the congruence of X. Then

$$\nabla_X \nabla_X Z = R(X, Z, X) \tag{14.8}$$

Proof. Since $[X, Z] = 0$ and $T = 0$, we have $\nabla_X Z = \nabla_Z X$. Then

$$
\begin{aligned}
\nabla_X \nabla_X Z \quad &= \quad \nabla_X \nabla_Z X \\
&= \quad \nabla_Z \nabla_X X + R(X, Z, X) \\
&\overset{X \text{ is geodesic}}{=} \quad R(X, Z, X)
\end{aligned}
$$

□

Suppose α and β are two neighbouring integral curves of X. There exists $\epsilon > 0$ such that $\exp(\epsilon Z)$ maps points of α to points of β; equation (14.8) then determines how far apart α and β are.

In terms of components, if t is an affine parameter for the integral curves of X, equation (14.8) has the form

$$\frac{d^2 Z^a}{dt^2} = R^a{}_{bcd} X^b X^c Z^d$$

This gives us another geometric interpretation of the Riemann tensor.

14.13 Theorem (Bianchi's identities)

If $T^i_{jk} = 0$, then

$$R^i{}_{j[kl;m]} = 0 \tag{14.9}$$

Proof. We choose a normal coordinate system with respect to a point p. Then $\Gamma^a{}_{(bc)} = 0$ at p. Since $T^i_{jk} = 0$, we have $\Gamma^a{}_{[bc]} = 0$ and therefore $\Gamma^a{}_{bc} = 0$ at p. Then

$$R^i{}_{jkl;m} = \Gamma^i{}_{lj,\,km} - \Gamma^i{}_{kj,\,lm}$$

at p. Antisymmetrising, we obtain equation (14.9).

Since this is a tensor equation, it is valid in any coordinate system. □

Equations (14.9) are called the 'Bianchi's identities'.

14.14 Definition (Ricci tensor)

The quantities

$$R_{bd} \overset{\text{def}}{=} R^a{}_{bad} \tag{14.10}$$

are the components of a rank-$\begin{pmatrix} 0 \\ 2 \end{pmatrix}$ tensor called the 'Ricci tensor'.

14.15 Remark (more on curvature)

If $\dim(M) = n$, then R^i_{jkl} has n^4 components. For $n = 4$, $n^4 = 256$. Because of the symmetries of R^i_{jkl} and Bianchi's identities, only 20 of those 256 components are really independent (see exercise 15.10).

The Ricci tensor, R_{ij}, is symmetric, so it has ten independent components (out of 16 total components). We may think of R_{ij} as representing 10 of the 20 components of the Riemann tensor. The latter's other ten components are represented by the 'Weyl tensor', W^i_{jkl}, which has the same form as the Riemann tensor except for the fact that its corresponding 'Ricci tensor' vanishes: $W^i_{jil} = 0$.

Additionally, contracting on the Ricci tensor's indices we obtain the 'Ricci scalar', $R = R^i_i$. To do this, we require a metric, so we shall do it in chapter 15.

  (i) $\underline{n = 0}$: M is a point and has no curvature.
 (ii) $\underline{n = 1}$: M is a line and only has extrinsic curvature, as there are no areas around which we can carry out a parallel transport. The extrinsic curvature is determined by the local curvature radius and is the function κ studied in chapter 2 and in section 14.8.
(iii) $\underline{n = 2}$: A vector parallel-transported around a parallelogram on M remains in $T(M)$ at all times, and, since the parallel transport does not change its magnitude, the only change will be given by the angle between the original vector and the transported one (said angle does not depend on the particular choice of parallelogram, only on its area). Therefore, all the information about the intrinsic curvature of a two-dimensional manifold is contained in the Ricci scalar R (it is the Gaussian curvature). The four components of R_{ij} and the 16 components of R^i_{jkl} can be calculated from R.
 (iv) $\underline{n = 3}$: We may move a small circle along geodesics perpendicular to the circle's plane and see how it changes. We need six independent scalars to describe the circle's deformation: the change and the rate of change in three independent directions. R_{ij} is symmetric, so only six of its nine elements are independent. The 81 elements of R^i_{jkl} can be calculated from R_{ij}.
  (v) $\underline{n = 4}$: To do something in four dimensions analogous to the transport of the circle in three dimensions, we would have to take a sphere and transport it along geodesics into the fourth dimension. The sphere will change both in volume and in shape; the ten independent components of R_{ij} describe how its volume changes in each direction. The ten components of W^i_{jkl} describe how its shape changes. $W^i_{jkl} = 0$ and $R_{ij} \neq 0$ imply a change in volume but not in shape; $W^i_{jkl} \neq 0$ and $R_{ij} = 0$ imply a change in shape maintaining its volume. One can therefore speak of 'Ricci curvature' and 'Weyl curvature'. The complete curvature is a combination of both and is given by the Riemann tensor.

Chapter 15

Pseudo-Riemannian metric

So far, we have introduced and exploited several constructions on a manifold M and its tangent bundle $T(M)$, but we have not mentioned 'inner products'. Having an inner product would give us a way to measure, that is, a 'metric', and, as we shall see, having a metric naturally selects a connection. This makes the calculation of curvature easier, as will be illustrated by a few examples, and allows us to obtain the geodesic equation from a variational principle.

15.1 Definition (inner product)

The rank-$\begin{pmatrix} 0 \\ 2 \end{pmatrix}$ tensor $\langle \, , \, \rangle_p : T_p(M) \times T_p(M) \longrightarrow \mathbb{R}$ on M is called an 'inner product' on $T_p(M)$ if and only if the following is true:
  (i) $\langle X, Y \rangle_p = \langle Y, X \rangle_p$ for all $X, Y \in T_p(M)$ (i.e. it is symmetric),
  (ii) if for all $Y \in T_p(M)$ we have $\langle X, Y \rangle_p = 0$, then $X = 0$ (i.e. it is nondegenerate).

15.2 Example

Let $M = \mathbb{R}^n$. For each $r \in \{1, \ldots, n\}$ we define an inner product $\langle \, , \, \rangle_r$ as follows:

$$\langle a, b \rangle_r = \sum_{i=1}^{r} a^i b^i - \sum_{i=r+1}^{n} a^i b^i$$

(when $r = n$, we obtain the usual inner product in $\mathbb{R}^n$).
  Note that, for any given r, $\langle \, , \, \rangle_r$ is nondegenerate: if $a \neq 0$, we have

$$\langle (a^1, \ldots, a^n), (a^1, \ldots, a^r, -a^{r+1}, \ldots, -a^n) \rangle_r = \sum_{i=1}^{n} (a^i)^2 > 0$$

doi:10.1088/2053-2563/aadf65ch15

15.3 Remark

Let $\{e_a\}$ be a basis of $T_p(M)$ and $\{\omega^a\}$ be its dual basis in $T_p^*(M)$. Then we may write

$$\langle \ , \ \rangle_p = g_{ij}\omega^i \otimes \omega^j \tag{15.1}$$

where

$$g_{ij} = \langle e_i, e_j \rangle_p \tag{15.2}$$

The symmetry of $\langle \ , \ \rangle_p$ implies that $g_{ij} = g_{ji}$. Its nondegeneracy implies that $\det(g_{ij}) \neq 0$, that is, (g_{ij}) is nonsingular.

We define

$$(g^{ij}) \overset{\text{def}}{=} (g_{ij})^{-1} \tag{15.3}$$

that is, $g^{ij}g_{jk} = \delta^i{}_k$.

15.4 Definition (magnitude, angle, and length)

Let $p \in M$; $X, \ Y \in \mathfrak{X}(M)$; and γ be an integral curve of X passing through p.

(i) $\|X_p\| = \langle X_p, X_p \rangle_p^{\frac{1}{2}}$ is called the 'magnitude' of X_p.

(ii) $\arccos\left(\frac{\langle X_p, Y_p \rangle_p}{\| X_p \| \| Y_p \|}\right)$ is called the 'angle' between X_p and Y_p.

(iii) $l(\gamma)_{t_1}^{t_2} \overset{\text{def}}{=} \int_{t_1}^{t_2} \|X\| \, dt$ is called the 'length of γ from t_1 to t_2'.

15.5 Theorem (properties of the magnitude)

Let $X, Y \in T_p(M)$, $a \in \mathbb{R}$, and $\langle \ , \ \rangle_p$ be strictly positive. Then

(i) $\|aX\| = |a| \, \|X\|$.

(ii) $\left|\langle X, Y \rangle_p\right| \leqslant \|X\| \, \|Y\|$ (with equality iff X and Y are linearly dependent).

(iii) $\|X + Y\| \leqslant \|X\| + \|Y\|$.

Proof. Since (i) is trivial, we shall only prove (ii) and (iii).

(ii) If X and Y are linearly dependent, the equality is a consequence of (i). Otherwise, $sX - Y \neq 0$ for all $s \in \mathbb{R}$; thus,

$$0 < \|sX - Y\|^2 = \langle sX - Y, sX - Y \rangle_p$$
$$= s^2\|X\|^2 - 2s\langle X, Y \rangle_p + \|Y\|^2$$

The right-hand side of the equation is a quadratic polynomial in s and has no real solutions (because there is no s for which $Y = s \, X$), so its discriminant is negative:

$$4\langle X, Y \rangle_p^2 - 4\|X\|^2\|Y\|^2 < 0$$

that is,

$$\langle X, Y \rangle_p < \|X\| \, \|Y\|$$

(iii) We write

$$\|X + Y\|^2 = \langle X + Y, X + Y \rangle_p$$
$$= \|X\|^2 + \|Y\|^2 + 2\langle X, Y \rangle_p$$
$$\overset{\text{(ii)}}{\leqslant} \|X\|^2 + \|Y\|^2 + 2\|X\| \, \|Y\|$$
$$= (\|X\| + \|Y\|)^2$$

□

15.6 Definition (metric)

Let M be an n-dimensional differentiable manifold and $\mathcal{T}^{(0,2)}$ be the space of rank-$\binom{0}{2}$ tensor fields on M. Then

(i) The map

$$\langle \, , \, \rangle : M \longrightarrow \mathcal{T}^{(0,2)}$$
$$p \longmapsto \langle \, , \, \rangle_p$$

is called a 'metric' on M.

(ii) The quantities g_{ij} defined above are called the 'metric coefficients'. We shall denote the metric by (g_{ij}).

(iii) $\text{sgn}(g_{ij}) \overset{\text{def}}{=} \begin{pmatrix} \text{number of positive} \\ \text{eigenvalues of } (g_{ij}) \end{pmatrix} - \begin{pmatrix} \text{number of negative} \\ \text{eigenvalues of } (g_{ij}) \end{pmatrix}$ is called the 'signature of g_{ij}'.

(iv) If $|\text{sgn}(g_{ij})| = n$ (that is, $\langle \, , \, \rangle$ is positive definite), the metric is called a 'Riemannian metric' and (M, g_{ij}) is called a 'Riemannian manifold'.
If $|\text{sgn}(g_{ij})| = n - 2$, the metric is called a 'Lorentzian' or 'pseudo-Riemannian metric' and (M, g_{ij}) is called a 'Lorentzian' or 'pseudo-Riemannian manifold'.

Suppose $n = 4$. From elementary linear algebra, we know that, at a point p, we can always express a Lorentzian metric as

$$g_{ij} = \text{diag}(-1, 1, 1, 1)$$

by using a coordinate transformation.

A vector $X \in T_p(M)$ is called 'time-like' if $g_{ij}X^iX^j < 0$, 'null' if if $g_{ij}X^iX^j = 0$ and 'space-like' if $g_{ij}X^iX^j > 0$.

15.7 Remark (metric $\Rightarrow T_p(M) \cong T_p^*(M)$)

Since (g_{ab}) is nonsingular, there exists a non-natural isomorphism between $T_p(M)$ and its dual space, $T_p^*(M)$, given by

$$X^a = g^{ab}\eta_b \quad \text{for} \quad \eta \in T_p^*(M)$$

$$\eta_a = g_{ab}X^b \quad \text{for} \quad X \in T_p(M)$$

We can then identify $X \longleftrightarrow \eta$ and denote by X both the vector and the 1-form. This isomorphism is also valid for tensors:

$$T^{ab} \longleftrightarrow T^a{}_b \longleftrightarrow T_a{}^b \longleftrightarrow T_{ab}$$

where the indices are 'raised' and 'lowered' via g^{ab} and g_{ab} as in the case of vectors and 1-forms, that is,

$$T^a{}_b = g^{ac}T_{cb}$$
$$T_{ab} = g_{bc}T_a{}^c$$
$$\text{etc}$$

From now on, we shall consider that tensors identified with each other in this way are different representations of the same object.

15.8 Metric connection

For now, let us write $\langle \, , \, \rangle = g$, that is, $g = (g_{ij})$. Suppose we have a connection ∇ with the property that

$$\nabla g = 0$$

Let X, Y, and Z be three vector fields on M. Since $g(Y, Z)$ is a function, we have

$$
\begin{aligned}
X[g(Y, Z)] \ &= \ \nabla_X[g(Y, Z)] \\
&\overset{\text{Leibnitz}}{=} (\nabla_X g)(Y, Z) + g(\nabla_X Y, Z) + g(Y, \nabla_X Z) \\
&\overset{\nabla g=0}{=} g(\nabla_X Y, Z) + g(Y, \nabla_X Z)
\end{aligned}
$$

Analogously,

$$Y[g(Z, X)] = g(\nabla_Y Z, X) + g(Z, \nabla_Y X)$$
$$Z[g(X, Y)] = g(\nabla_Z X, Y) + g(X, \nabla_Z Y)$$

Adding the first two of these equations and subtracting the third one,

$$
\begin{aligned}
X[g(Y, Z)] + Y[g(Z, X)] - Z[g(X, Y)] &= g(Z, \nabla_X Y) + g(Y, \nabla_X Z) + g(\nabla_Y Z, X) \\
&\quad + g(Z, \nabla_Y X) - g(\nabla_Z X, Y) \\
&\quad - g(X, \nabla_Z Y) \\
&= 2g(Z, \nabla_X Y) + g(Y, [X, Z]) \\
&\quad + g(X, [Y, Z]) + g(Z, [Y, X])
\end{aligned}
$$

where the second equality results from the fact that $\nabla_X Y - \nabla_Y X = [X, Y]$. Put differently,

$$g(Z, \nabla_X Y) = \frac{1}{2}[-Z[g(X, Y)] + Y[g(Z, X)] + X[g(Y, Z)]$$
$$+ g(Z, [X, Y]) + g(Y, [Z, X]) - g(X, [Y, Z])]$$

Let $\{e_a\}$ be a basis of $\mathfrak{X}(M)$, and let us set $Z = e_a$, $X = e_b$, $Y = e_c$. Then

$$g(e_a, \nabla_b e_c) = \frac{1}{2}(e_c\, g_{ab} + e_b\, g_{ca} - e_a\, g_{bc} + \gamma^d_{bc}\, g_{ad} + \gamma^d_{ab}\, g_{cd} - \gamma^d_{ca}\, g_{bd})$$

But

$$g(e_a, \nabla_b e_c) = g(e_a, \Gamma^d_{cb} e_d) = \Gamma^d_{cb}\, g_{ad}$$

If the basis is coordinate, we have

$$\Gamma^i_{jk} = \frac{1}{2}g^{im}(g_{km,\,j} + g_{jm,\,k} - g_{jk,\,m})$$

that is, the connection components are completely determined by the metric.

15.9 Theorem (uniqueness of the metric connection)

Let (M, g) be a manifold with a metric. Then there exists a unique connection such that

$$\nabla g = 0 \tag{15.4}$$

In a coordinate basis, the components of this connection are

$$\Gamma^i_{jk} = \frac{1}{2}g^{im}(g_{km,\,j} + g_{jm,\,k} - g_{jk,\,m}) \tag{15.5}$$

This connection is called the 'metric connection' or 'Levi-Civita connection', and its components Γ^i_{jk} are called the 'Christoffel symbols'.

Proof. See remark 15.8. □

Under the assumption that there is no torsion, we have proven that there is a unique connection satisfying $\nabla g = 0$. We can also see that, if ∇ is the Levi-Civita connection, then M has no torsion.

General relativity assumes that the connection is metric, so from now on we shall use such a connection unless explicitly stated otherwise.

15.10 Exercise (symmetries of the Riemann tensor)

We define $R_{abcd} = g_{af}R^f_{bcd}$. From the definition of the Riemann tensor, we know that $R^a_{b(cd)} = 0$. From equation (14.4), we know that $R^a_{[bcd]} = 0$.

Show that, if the connection is metric, the Riemann tensor also has the following symmetries:

$$R_{(ab)cd} = 0 \tag{15.6}$$

$$R_{abcd} = R_{cdab} \tag{15.7}$$

Solution. In normal coordinates, from equations (14.4) and (15.5) we obtain

$$R_{ijkl} = \frac{1}{2}(g_{il,\,jk} + g_{kj,\,il} - g_{lj,\,ik} - g_{ik,\,jl}) \tag{15.8}$$

from which equations (15.6) and (15.7) follow immediately.

15.11 Definition (Einstein tensor and Ricci scalar)

The 'Ricci scalar' is defined as $R \overset{\text{def}}{=} g^{ab}R_{ab}$. The 'Einstein tensor' is defined as $G_{ab} \overset{\text{def}}{=} R_{ab} - \frac{1}{2}R\,g_{ab}$.

15.12 Exercise

Show that the contraction of Bianchi's identities is equivalent to

$$\nabla_a G{}^a{}_b = 0 \tag{15.9}$$

Solution. Left to the reader.

15.13 Exercise (light cone)

In the 'Minkowski space', the space–time of special relativity, the metric has the form

$$g_{ij} = \text{diag}(-1,\, 1,\, 1,\, 1)$$

Show that the set of null vectors at a point forms a double cone with its vertex at the point in question. This cone is called the 'light cone' and separates the time-like vectors from the space-like ones.

Deduce that in a four-dimensional Lorentzian manifold the same structure exists for the tangent space at each point on the manifold.

Solution. If X is a null vector, then $0 = g(X, X) = (X^1)^2 + (X^2)^2 + (X^3)^2 - (X^0)^2$, that is, $(X^0)^2 = (X^1)^2 + (X^2)^2 + (X^3)^2$, which is the equation of a cone. If Y is a time-like vector, then $(Y^1)^2 + (Y^2)^2 + (Y^3)^2 - (Y^0)^2 < 0$, that is, Y lies within the cone. If Z is a space-like vector, then $(Z^1)^2 + (Z^2)^2 + (Z^3)^2 - (Z^0)^2 > 0$, that is, Z lies outside the cone. We know that at any point we may write a Lorentzian metric (using a coordinate transformation) as $(g_{ij}) = \text{diag}(-1, 1, 1, 1)$, so the tangent space has the same structure at each point on the manifold.

15.14 Definition (orthogonality)

$X, Y \in T_p(M)$ are called 'orthogonal' (to each other) if and only if

$$\langle X, Y \rangle_p = 0$$

15.15 Theorem (*Riemann = 0 ⇔ M* **is flat**)

The Riemann tensor vanishes if and only if the manifold is flat.

Proof. '⇐'

In a flat space the geodesics are straight lines, i.e. there exists a coordinate system $\{x^i\}_{i=1,\,...,\,n}$ in which

$$\ddot{x}^i = 0$$

From equation (13.5) for a geodesic we then have $\Gamma^i_{jk} \equiv 0$, and from the expression (14.4) of the Riemann tensor in terms of the connection coefficients it follows

$$R^i{}_{jk\ell} = 0$$

'⇒'

Let $\{x^i\}_{i=1,\,...,\,n}$ be a local coordinate system in M. Let $p \in M$ be an arbitrary point in M and, without loss of generality, let $\dim(M) = 4$.

Let us take, in $T_p(M)$, an orthonormal set of vectors $(T, X, Y, Z)_p$ with

$$g_{ij}T^i_p T^j_p = -1, \ g_{ij}X^i_p X^j_p = 1, \ g_{ij}Y^i_p Y^j_p = 1, \ g_{ij}Z^i_p Z^j_p = 1.$$

Since $R_{ijk\ell} = 0$, parallel transport of these vectors around a closed curve does not change them. We may then use this parallel transport to define vector fields (T, X, Y, Z) in a neighbourhood U of p, by simply transporting $(T, X, Y, Z)_p$ away from p in such a way that

$$\nabla_T T = 0, \ \nabla_X X = 0, \ \nabla_Y Y = 0, \ \nabla_Z Z = 0$$

This is equivalent to

$$\frac{\partial T^i}{\partial x^k} = -\Gamma^i{}_{jk}T^j, \qquad \frac{\partial X^i}{\partial x^k} = -\Gamma^i{}_{jk}X^j$$

$$\frac{\partial Y^i}{\partial x^k} = -\Gamma^i{}_{jk}Y^j, \qquad \frac{\partial Z^i}{\partial x^k} = -\Gamma^i{}_{jk}Z^j$$

and since $\Gamma^i{}_{jk} = \Gamma^i{}_{kj}$ (assuming that the torsion in M vanishes), then

$$\frac{\partial T^i}{\partial x^k} = \frac{\partial T^k}{\partial x^i}, \ ... \text{ (idem for all other vectors),}$$

i.e. each of these fields is irrotational, so we may write it as a gradient of a scalar field

$$A^{(a)} \overset{\text{def}}{=} (T, X, Y, Z) = \nabla\Phi^{(a)}$$

On the other hand, parallel transport preserves inner products, so that

$$g^{ik}\frac{\partial\Phi^{(a)}}{\partial x^i}\frac{\partial\Phi^{(b)}}{\partial x^k} = \eta^{ab}$$

with $\eta^{ab} = \mathrm{diag}(-1, 1, 1, 1)$. This means that there exists a coordinate transcription, namely $x'^k = \Phi^{(k)} x^k$, which sends (U, g_{ik}) to *Minkowski space* (U, η_{ab}) (which will be studied later on), which is flat and in which the geodesics are straight lines. Using the same procedure we can extend this to all of M. □

15.16 Example (curvature of the sphere $\mathbb{S}^2$)

Let us calculate the components of the Riemann tensor for the sphere $\mathbb{S}^2$ with the metric $ds^2 = r^2(d\theta^2 + \sin^2\theta\, d\phi^2)$:

From remark 14.15, for $n = 2$ we know that there is a single independent component of ${}^i_{jkl}$. We select

$$R_{\theta\phi\theta\phi} = g_{\theta\theta} R^\theta{}_{\phi\theta\phi} \overset{\underset{\text{lemma 14.7}}{\text{equation (14.4)}}}{=} g_{\theta\theta}(\Gamma^\theta{}_{\phi\phi,\theta} - \Gamma^\theta{}_{\theta\phi,\phi} + \Gamma^\theta{}_{\theta\alpha}\Gamma^\alpha{}_{\phi\phi} - \Gamma^\theta{}_{\phi\alpha}\Gamma^\alpha{}_{\phi\theta})$$

In our case, we have

$$g_{\theta\theta} = r^2$$
$$g_{\phi\phi} = r^2 \sin^2\theta$$
$$g_{\alpha\beta} = 0 \quad \text{in all other cases}$$

Therefore,

$$g^{\theta\theta} = \frac{1}{r^2}$$
$$g^{\phi\phi} = \frac{1}{r^2 \sin^2\theta}$$
$$g^{\alpha\beta} = 0 \quad \text{in all other cases}$$

From equation (15.5), we know that

$$\Gamma^i{}_{jk} = \frac{1}{2}g^{im}(g_{mk,j} + g_{jm,k} - g_{jk,m})$$

whereby

$$\Gamma^\theta{}_{\phi\phi} = \frac{1}{2}g^{\theta\theta}(-g_{\phi\phi,\theta}) = -\frac{1}{2}\frac{1}{r^2}2r^2 \sin\theta\cos\theta = -\sin\theta\cos\theta$$

$$\Gamma^\phi{}_{\theta\phi} = \frac{1}{2}g^{\phi\phi}(g_{\phi\phi,\theta}) = \frac{1}{2}\frac{1}{r^2\sin^2\theta}2r^2\sin\theta\cos\theta = \frac{\cos\theta}{\sin\theta} = \cot\theta$$

$$\Gamma^i{}_{jk} = 0 \quad \text{in all other cases}$$

Therefore,

$$\begin{aligned}
R_{\theta\phi\theta\phi} &= g_{\theta\theta}(\Gamma^\theta{}_{\phi\phi,\theta} - \Gamma^\theta{}_{\phi\phi}\Gamma^\phi{}_{\phi\theta})\\
&= r^2(-\cos^2\theta + \sin^2\theta + \sin\theta\cos\theta\cot\theta)\\
&= r^2 \sin^2\theta
\end{aligned}$$

Put differently,

$$r^2 \sin^2 \theta = R_{\theta\phi\theta\phi} = R_{\phi\theta\phi\theta} = -R_{\theta\phi\phi\theta} = -R_{\phi\theta\theta\phi}$$

Calculating the Ricci tensor, we have

$$R_{\theta\theta} = R^a{}_{\theta a\theta} = R^\phi{}_{\theta\phi\theta} = g^{\phi\phi} R_{\phi\theta\phi\theta} = \frac{r^2 \sin^2 \theta}{r^2 \sin^2 \theta} = 1$$

$$R_{\phi\phi} = R^a{}_{\phi a\phi} = R^\theta{}_{\phi\theta\phi} = g^{\theta\theta} R_{\theta\phi\theta\phi} = \frac{r^2 \sin^2 \theta}{r^2} = \sin^2 \theta$$

$$R_{\theta\phi} = R^a{}_{\theta a\phi} = 0$$

The Ricci scalar is given by

$$R = g^{ab} R_{ab} = g^{\theta\theta} R_{\theta\theta} + g^{\phi\phi} R_{\phi\phi} = \frac{1}{r^2} + \frac{\sin^2 \theta}{r^2 \sin^2 \theta} = \frac{2}{r^2}$$

In two dimensions, the scalar curvature R is equal to twice the Gaussian curvature. For a surface in $\mathbb{E}^3$, we have

$$R = \frac{2}{\rho_1 \rho_2}$$

where ρ_1 and ρ_2 are the 'principal radii' of the surface, that is, the radii of the curves marked on the surface by a plane perpendicular to it in the principal directions (see chapter 2).

15.17 Example (metric of the n-sphere $\mathbb{S}^n$)

Note that for $\mathbb{S}^2$ (seen as though submerged in $\mathbb{R}^3$) we have

$$z = r \cos \theta$$
$$x = r \sin \theta \cos \phi$$
$$y = r \sin \theta \sin \phi$$

Then

$$\frac{\partial}{\partial \theta} = r(\cos \theta \cos \phi, \, \cos \theta \sin \phi, \, -\sin \theta)$$

$$\frac{\partial}{\partial \phi} = r(-\sin \theta \sin \phi, \, \sin \theta \cos \phi, \, 0)$$

whereby

$$g_{\theta\theta} = \left\langle \frac{\partial}{\partial \theta}, \frac{\partial}{\partial \theta} \right\rangle = r^2$$

$$g_{\theta\phi} = \left\langle \frac{\partial}{\partial \theta}, \frac{\partial}{\partial \phi} \right\rangle = 0$$

$$g_{\phi\phi} = \left\langle \frac{\partial}{\partial \phi}, \frac{\partial}{\partial \phi} \right\rangle = r^2 \sin^2 \theta$$

that is,

$$(g_{ij}) = \begin{pmatrix} r^2 & 0 \\ 0 & r^2 \sin^2 \theta \end{pmatrix}$$

For the n-dimensional case, let us denote by $(x^1, \ldots, x^{n+1})$ the coordinates of $\mathbb{R}^{n+1}$, and by $(u^1, \ldots, u^n)$ the angular coordinates of $\mathbb{S}^n$. Then

$$x^1 = r \cos u^1$$
$$x^2 = r \sin u^1 \cos u^2$$
$$x^3 = r \sin u^1 \sin u^2 \cos u^3$$
$$\cdots$$
$$x^n = r \sin u^1 \sin u^2 \cdots \sin u^{n-1} \cos u^n$$
$$x^{n+1} = r \sin u^1 \sin u^2 \cdots \sin u^{n-1} \sin u^n$$

whereby

$$\frac{\partial}{\partial u^1} = r(-\sin u^1, \cos u^1 \cos u^2, \cos u^1 \sin u^1 \cos u^3, \ldots ,$$
$$\cos u^1 \sin u^2 \cdots \sin u^{n-1} \cos u^n, \cos u^1 \sin u^2 \cdots \sin u^{n-1} \sin u^n)$$

$$\frac{\partial}{\partial u^2} = r(0, -\sin u^1 \sin u^2, \ldots , \sin u^1 \cos u^2 \sin u^3 \cdots \sin u^{n-1} \cos u^n,$$
$$\sin u^1 \cos u^2 \sin u^3 \cdots \sin u^{n-1} \sin u^n)$$

$$\frac{\partial}{\partial u^3} = r(0, 0, -\sin u^1 \sin u^2 \sin u^3, \sin u^1 \sin u^2 \cos u^3 \cos u^4, \ldots ,$$
$$\sin u^1 \sin u^2 \cos u^3 \sin u^4 \cdots \sin u^{n-1} \cos u^n,$$
$$\sin u^1 \sin u^2 \cos u^3 \sin u^4 \cdots \sin u^{n-1} \sin u^n)$$
$$\cdots$$

Therefore,

$$g_{11} = \left\langle \frac{\partial}{\partial u^1}, \frac{\partial}{\partial u^1} \right\rangle = r^2$$

$$g_{22} = \left\langle \frac{\partial}{\partial u^2}, \frac{\partial}{\partial u^2} \right\rangle = r^2 \sin^2 u^1$$

$$g_{33} = \left\langle \frac{\partial}{\partial u^3}, \frac{\partial}{\partial u^3} \right\rangle = r^2 \sin^2 u^1 \sin^2 u^2$$
$$\cdots$$

$$g_{nn} = \left\langle \frac{\partial}{\partial u^n}, \frac{\partial}{\partial u^n} \right\rangle = r^2 \sin^2 u^1 \sin^2 u^2 \cdots \sin^2 u^{n-1}$$

$$g_{ij} = 0 \quad \text{for } i \neq j$$

that is,

$$(g_{ij}) = \begin{pmatrix} r^2 & 0 & 0 & \cdots & 0 \\ 0 & r^2 \sin^2 u^1 & 0 & \cdots & 0 \\ 0 & 0 & r^2 \sin^2 u^1 \sin^2 u^2 & \cdots & 0 \\ \vdots & \vdots & \vdots & \ddots & \vdots \\ 0 & 0 & 0 & \cdots & r^2 \sin^2 u^1 \cdots \sin^2 u^{n-1} \end{pmatrix}$$

From this metric we can calculate the Christoffel symbols, the Riemann tensor, and so on.

15.18 Example (curvature of the torus $\mathbb{T}^2$)

For a torus with radii $R = b$ and $r = a$, we establish a coordinate system (θ, ϕ) as shown in figure 15.1.

Note that (orthogonal) displacements in ϕ and θ have lengths $a\, d\phi$ and $[b + a \cos \phi] d\theta$, respectively, that is,

$$g_{\theta\theta} = [b + a \cos \phi]^2$$
$$g_{\phi\phi} = a^2$$
$$g_{\theta\phi} = g_{\phi\theta} = 0$$

The only nonzero Christoffel symbols are

$$\Gamma^{\phi}{}_{\theta\theta} = \frac{1}{a}[b + a \cos \phi] \sin \phi$$

$$\Gamma^{\theta}{}_{\phi\theta} = \Gamma^{\theta}{}_{\theta\phi} = -\frac{a \sin \phi}{b + a \cos \phi}$$

whereby

$$R_{\phi\theta\phi\theta} = a \cos \phi [b + a \cos \phi]$$

and, as before,

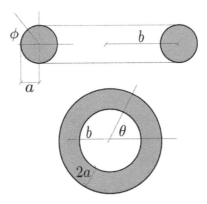

Figure 15.1. Coordinate system for the torus $\mathbb{T}^2$.

$$R_{\phi\theta\phi\theta} = R_{\theta\phi\theta\phi} = -R_{\phi\theta\theta\phi} = -R_{\theta\phi\phi\theta}$$

Calculating the Ricci tensor, we have

$$R_{\theta\theta} = R^{c}_{\ \theta c\theta} = R^{\phi}_{\ \theta\phi\theta} = g^{\phi\phi}R_{\phi\theta\phi\theta} = \frac{1}{a}\cos\phi[b + a\cos\phi]$$

$$R_{\phi\phi} = R^{c}_{\ \phi c\phi} = R^{\theta}_{\ \phi\theta\phi} = g^{\theta\theta}R_{\theta\phi\theta\phi} = \frac{a\cos\phi}{b + a\cos\phi}$$

$$R_{\theta\phi} = R^{c}_{\ \theta c\phi} = 0 = R_{\phi\theta}$$

The Ricci scalar is given by

$$\begin{aligned}
R &= g^{ab}R_{ab} \\
&= g^{\theta\theta}R_{\theta\theta} + g^{\phi\phi}R_{\phi\phi} \\
&= \frac{\cos\phi[b + a\cos\phi]}{a[b + a\cos\phi]^2} + \frac{a\cos\phi}{a^2[b + a\cos\phi]} \\
&= \frac{\cos\phi}{a[b + a\cos\phi]} + \frac{\cos\phi}{a[b + a\cos\phi]} \\
&= \frac{2\cos\phi}{a[b + a\cos\phi]}
\end{aligned}$$

Note the following:
    (i) $R \neq R(\theta)$.
    (ii) If $\phi = \pm\frac{\pi}{2}$, then $R = 0$.
    (iii) If $\phi \in \left(-\frac{\pi}{2}, \frac{\pi}{2}\right)$, then $\cos\phi > 0$ and thus $R > 0$.
    (iv) If $\phi \in \left(\frac{\pi}{2}, \frac{3\pi}{2}\right)$, then $\cos\phi < 0$ and thus $R < 0$ (for $b > a$).

15.19 Example (geodesics of the torus)

We have obtained the metric of the torus $\mathbb{T}^2$ by geometric methods. Another way to do it is by calculating the coefficients E, F, and G of the first fundamental form, as we saw in chapter 2. The torus submerged in $\mathbb{R}^3$ is parametrised by (see chapter 1 or chapter 2)

$$(\theta, \phi) \longrightarrow (x^1, x^2, x^3)$$

where

$$\begin{aligned}
x^1 &= [b + a\cos\phi]\cos\theta \\
x^2 &= [b + a\cos\phi]\sin\phi \\
x^3 &= a\sin\phi
\end{aligned} \tag{15.10}$$

and ϕ and θ are, respectively, the polar and azimuthal angles, as in the previous exercise. It is shown in figure 15.2.

Figure 15.2. Parametrised torus in $\mathbb{R}^3$.

Hence,

$$x_\theta = (-[b + a \cos \phi] \sin \theta, [b + a \cos \phi] \cos \theta, 0)$$
$$x_\phi = (-a \cos \theta \sin \phi, -a \sin \theta \sin \phi, a \cos \theta)$$

and thus

$$E = x_\theta \cdot x_\theta = (b + a \cos \phi)^2$$
$$F = x_\theta \cdot x_\phi = 0$$
$$G = x_\phi \cdot x_\phi = a^2 \sin^2 \phi + a^2 \cos^2 \phi = a^2$$

that is,

$$(g_{ij}) = \begin{pmatrix} [b + a \cos \phi]^2 & 0 \\ 0 & a^2 \end{pmatrix}$$

$$(g^{ij}) = \begin{pmatrix} \dfrac{1}{[b + a \cos \phi]^2} & 0 \\ 0 & \dfrac{1}{a^2} \end{pmatrix}$$

We calculate the Christoffel symbols from the metric:

$$\Gamma^\theta{}_{\theta\phi} = \Gamma^\theta{}_{\phi\theta} = -\frac{a \sin \phi}{b + a \cos \phi}$$

$$\Gamma^\phi{}_{\theta\theta} = \frac{1}{a}[b + a \cos \phi] \sin \phi$$

and the others are zero. The equation for the geodesics is

$$\ddot{x}^a + \Gamma^a{}_{bc} \dot{x}^b \dot{x}^c = 0$$

whereby

$$\ddot{\theta} - \frac{2a \sin \phi}{b + a \cos \phi} \dot{\theta}\dot{\phi} = 0 \tag{15.11}$$

$$\ddot{\phi} + \frac{1}{a} \sin \phi (b + a \cos \phi) \dot{\theta}^2 = 0 \tag{15.12}$$

Let $u = b + a \cos \phi$. Then $\dot{u} = -a \sin \phi \dot{\phi}$ and equation (15.11) becomes

$$\frac{\ddot{\theta}}{\dot{\theta}} = -2 \frac{\dot{u}}{u}$$

Integrating,

$$\int \frac{\ddot{\theta}}{\dot{\theta}} = -2 \int \frac{\dot{u}}{u}$$

that is,

$$\begin{aligned}
\ln(\dot{\theta}) &= -2 \ln(u) + \ln(k) \\
&= \ln(u^{-2}) + \ln(k) \\
&= \ln(ku^{-2})
\end{aligned}$$

(where k is an integration constant), whereby

$$\dot{\theta} = ku^{-2} = \frac{k}{u^2} = \frac{k}{[b + a \cos \phi]^2} \tag{15.13}$$

Substituting equation (15.13) into equation (15.12),

$$\ddot{\phi} + \frac{k^2}{a[b + a \cos \phi]^3} \sin \phi = 0$$

But $-\frac{\dot{u}}{a} = \sin \phi \dot{\phi}$ and $u = b + a \cos \phi$, so

$$\ddot{\phi} \dot{\phi} + \frac{k^2}{au^3} \left(-\frac{\dot{u}}{a} \right) = 0$$

that is,

$$\ddot{\phi} \dot{\phi} = \frac{k^2 \dot{u}}{a^2 u^3} \tag{15.14}$$

Integrating,

$$\frac{1}{2} \dot{\phi}^2 = \int \ddot{\phi} \dot{\phi} = \frac{k^2}{a^2} \int \frac{\dot{u}}{u^3} = -\frac{k^2}{a^2} \frac{1}{2u^2} + \frac{1}{2} k' \tag{15.15}$$

(where k' is another integration constant), whereby

$$\dot{\phi}^2 = -\frac{k^2}{a^2 [b + a \cos \phi]^2} + k' \tag{15.16}$$

Equations (15.13) and (15.16) determine the geodesics on the torus $\mathbb{T}^2$, but we still have the freedom of the curves' parametrisation. Parametrisation by arc length gives

$$1 = ||\dot\gamma(t)||^2 = |g_{\theta\theta}\dot\theta^2 + g_{\phi\phi}\dot\phi^2| = \frac{k^2}{(b + a\cos\phi)^2} - \frac{k^2}{(b + a\cos\phi)^2} + a^2k' = a^2k'$$

that is,

$$k' = \frac{1}{a^2}$$

Substituting this into equation (15.16),

$$\dot\phi^2 = -\frac{k^2}{a^2(b + a\cos\phi)^2} + \frac{1}{a^2} = \frac{1}{a^2}\left[1 - \frac{k^2}{(b + a\cos\phi)^2}\right]$$

and

$$a^2\dot\phi^2 + k\dot\theta = 1 \tag{15.17}$$

Note that equations (15.13) and (15.17) determine the geodesics of $\mathbb{T}^2$ and have a single free parameter, k, which is appropriate because, given a point on $\mathbb{T}^2$, a geodesic passing through that point is completely determined by its direction.

It is usually assumed, erroneously, that the torus's geodesics are spirals with fixed step. This is because the geodesics on a plane are straight lines and a torus can be seen as $\mathbb{S}^1 \times \mathbb{S}^1$, that is, as the surface obtained by identifying opposite sides of a flat rectangle. The geodesics of this 'flat torus' are indeed the aforementioned spirals, but a torus with the parametrisation we have taken here, whose geometry is much richer, has nontrivial geodesics (as shown in figures 15.3 and 15.4 for different values of k).

15.20 Exercise

(i) Find the connection coefficients and the components of the Riemann tensor for the two-dimensional space

$$ds^2 = dv^2 - v^2 \, du^2$$

(ii) Show that, under the coordinate transformation

$$x = v\cosh u$$
$$t = v\sinh u$$

this space is the Minkowski space in two dimensions.

Solution.
(ii) Given the coordinate transformation above,

$$x^2 - t^2 = v^2$$
$$dx^2 = (dv\cosh u + du\, v\sinh u)^2$$
$$dt^2 = (dv\sinh u + du\, v\cosh u)^2$$

$k = 0$

$k = 0.5$

$k = 1$

$k = 1.5$

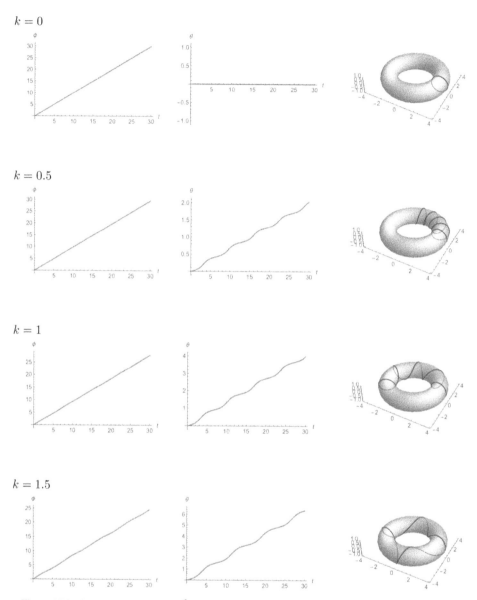

Figure 15.3. Geodesics of the torus $\mathbb{T}^2$ as parametrised by equations (15.10), for $k = 0, 0.5, 1, 1.5$.

Therefore,

$$dx^2 - dt^2 = dv^2 - v^2 du^2$$

(i) The only nonzero Christoffel symbols are

$$\Gamma^v_{uu} = v$$

$$\Gamma^u_{uv} = \Gamma^u_{vu} = \frac{1}{v}$$

15-16

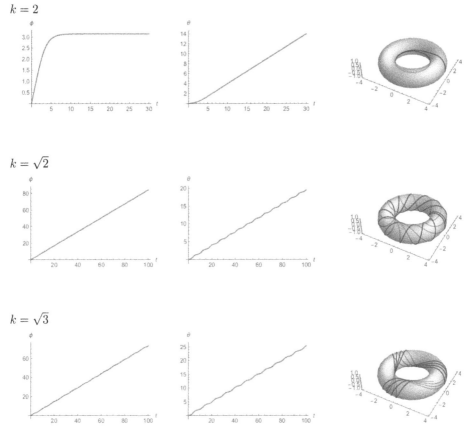

$k = 2$

$k = \sqrt{2}$

$k = \sqrt{3}$

Figure 15.4. Geodesics of the torus $\mathbb{T}^2$ as parametrised by equations (15.10) for $k = 2, \sqrt{2}, \sqrt{3}$.

Therefore,

$$R_{vuvu} = R^{v}_{uvu} = \Gamma^{v}_{uu,v} - \Gamma^{v}_{vu,u} + \Gamma^{v}_{v\alpha}\Gamma^{\alpha}_{uu} - \Gamma^{v}_{u\alpha}\Gamma^{\alpha}_{uv} = 1 - 0 + 0 - 1 = 0$$

Since there is only one independent component of the Riemann tensor in two dimensions, we have

$$R_{ijkl} \equiv 0$$

and the space is flat (which is obvious from (ii)).

15.21 Geodesic equation from the minimum-action principle

In chapter 13 we derived the equation for a geodesic from the fact that geodesics parallel-transport their tangent vectors. Here we shall derive it from a minimum-action principle.

In general relativity, the gravitational field is given by the curvature of space–time, a four-dimensional Lorentzian manifold (M, g_{ab}). As we have seen, this curvature is determined by the coefficients $\Gamma^a{}_{bc}$ of a connection ∇, and these (if the connection is metric) by the metric coefficients g_{ab}. The latter are then the field variables, and the Lagrangian for a particle in the presence of a gravitational field is purely kinetic.

In general, for a system with n degrees of freedom $\{q^a\}_{a\in\{1, \ldots, n\}}$, the most general kinetic Lagrangian is

$$L = \frac{1}{2}g_{ab}\,\dot{q}^a\dot{q}^b \tag{15.18}$$

with $g_{ab} = g_{ab}(q^c)$.

The action for this Lagrangian is $S = \int L\, dt$. The requirement that the action be minimum gives

$$
\begin{aligned}
0 \quad &= \quad \delta S \\
&= \quad \int \frac{1}{2}\left[\dot{q}^a\dot{q}^b\frac{\partial g_{ab}}{\partial q^i}\delta q^i + g_{ab}\dot{q}^a\delta\dot{q}^b + g_{ab}\dot{q}^b\delta\dot{q}^a\right]dt \\
&\overset{g_{ab}\text{ is symmetric}}{=} \int\left[\frac{1}{2}\dot{q}^a\dot{q}^b\frac{\partial g_{ab}}{\partial q^i}\delta q^i + g_{ab}\dot{q}^a\delta\dot{q}^b\right]dt \\
&= \quad \int\left[\frac{1}{2}\dot{q}^a\dot{q}^b\frac{\partial g_{ab}}{\partial q^i}\delta q^i + g_{ab}\dot{q}^a\frac{d}{dt}\delta q^b\right]dt \\
&= \quad \int\left[\frac{1}{2}\dot{q}^a\dot{q}^b\frac{\partial g_{ab}}{\partial q^i}\delta q^i + \frac{d}{dt}(g_{ab}\dot{q}^a\delta q^b) - \frac{d}{dt}(g_{ab}\dot{q}^a)\delta q^b\right]dt \\
&= \quad \int\left[\frac{1}{2}\dot{q}^a\dot{q}^b\frac{\partial g_{ab}}{\partial q^i}\delta q^i - \frac{d}{dt}(g_{ai}\dot{q}^a)\delta q^i\right]dt
\end{aligned}
$$

where the last equality results from the fact that $\delta q = 0$ at the integration limits.

For this equation to be satisfied by an arbitrary δq^i, it is necessary that

$$\frac{1}{2}\dot{q}^a\dot{q}^b\frac{\partial g_{ab}}{\partial q^i} - \frac{d}{dt}(g_{ai}\dot{q}^a) = 0$$

that is,

$$\frac{1}{2}\dot{q}^a\dot{q}^b\frac{\partial g_{ab}}{\partial q^i} - g_{ai}\frac{d}{dt}\frac{dq^a}{dt} - \dot{q}^a\frac{\partial g_{ai}}{\partial q^k}\dot{q}^k = 0 \tag{15.19}$$

Using the symmetry of g_{ab}, we write

$$\dot{q}^a\dot{q}^k\frac{\partial g_{ai}}{\partial q^i} = \frac{1}{2}\dot{q}^a\dot{q}^k\left(\frac{\partial g_{ai}}{\partial q^k} + \frac{\partial g_{ki}}{\partial q^a}\right) = \frac{1}{2}\dot{q}^a\dot{q}^b\left(\frac{\partial g_{ai}}{\partial q^b} + \frac{\partial g_{bi}}{\partial q^a}\right)$$

and thus, by virtue of equation (15.19),

$$\frac{1}{2}\dot{q}^a\dot{q}^b\left(\frac{\partial g_{ab}}{\partial q^i} - \frac{\partial g_{ai}}{\partial q^b} - \frac{\partial g_{bi}}{\partial q^a}\right) - g_{ai}\ddot{q}^a = 0$$

that is,

$$g_{ai}\ddot{q}^a + \frac{1}{2}\left(\frac{\partial g_{ai}}{\partial q^b} + \frac{\partial g_{bi}}{\partial q^a} - \frac{\partial g_{ab}}{\partial q^i}\right)\dot{q}^a\dot{q}^b = 0$$

Multiplying by g^{ic},

$$g^{ic}g_{ai}\ddot{q}^a + \frac{1}{2}g^{ic}\left(\frac{\partial g_{ai}}{\partial q^b} + \frac{\partial g_{bi}}{\partial q^a} - \frac{\partial g_{ab}}{\partial q^i}\right)\dot{q}^a\dot{q}^b = 0$$

The first term on the left is

$$g^{ic}g_{ai}\ddot{q}^a = \delta^c_a\ddot{q}^a = \ddot{q}^c$$

and the second one is

$$\Gamma^c_{ab}\dot{q}^a\dot{q}^b$$

Therefore,

$$\ddot{q}^c + \Gamma^c_{ab}\dot{q}^a\dot{q}^b = 0 \tag{15.20}$$

which is the geodesic equation obtained in theorem 13.2.

IOP Publishing

Differential Topology and Geometry with Applications to Physics

Eduardo Nahmad-Achar

Chapter 16

Newtonian space–time and thermodynamics

In the present chapter and the next we shall see how the language we have constructed allows for a greater mathematical simplicity and a more fundamental understanding of the physics we already know: Newtonian mechanics, thermodynamics, fluid dynamics, and electrodynamics.

16.1 Newtonian structure and Galilean coordinates

   (i) In Newtonian physics, space–time is a four-dimensional class-C^∞ differentiable manifold M in which there is an 'absolute time', that is, a function $t : M \longrightarrow \mathbb{R}$ such that $dt \not\equiv 0$.

  (ii) Two points $p, q \in M$ are called 'simultaneous' if and only if $t(p) = t(q)$.

 (iii) Each point $p \in M$ determines a set $S(p) = \{q \in M$ such that $t(p) = t(q)\}$, called a 'spatial section', which divides M in two nonintersecting regions: the 'past' and 'future' of p.

 (iv) On each $S(p)$ we have a Euclidean metric h such that $\text{sgn}(h) = 3$. We can choose global coordinates $\{x^i\}$ such that $h = \text{diag}(1, 1, 1)$.

  (v) The structure (M, t, h) is called a 'Galilean manifold'. A vector X is 'space-like' or 'time-like' depending on whether $dt(X) = 0$ or $dt(X) \neq 0$, respectively. Space-like vectors are the usual vectors of classical mechanics, while the integral curves of time-like vectors represent the history of moving particles.

 (vi) A 'Galilean coordinate system' is defined as follows (see figure 16.1): let γ be a time-like curve, and let $\theta \in \gamma$. We take at each point $q \in \gamma$ a set of linearly independent space-like vectors $\{e_\alpha\}_{\alpha \in \{1,2,3\}}$. We set $t(\theta) = 0$. Given an arbitrary point $p \in M$, let p_0 be the point where $S(p)$ intersects γ. As $\overrightarrow{p_0 p}$

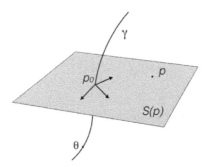

Figure 16.1. Construction of a Galilean coordinate system.

is space-like, we can write $\overrightarrow{p_0 p} = x_p^\alpha e_\alpha$. The Galilean coordinates of p are then

$$x^a(p) = \left(t(p_0),\ x_p^\alpha \right)$$

16.2 Transformation between reference systems

(i) Two Galilean systems of reference $F = (\gamma,\ \theta,\ \{e_\alpha\})$ and $F' = (\gamma',\ \theta',\ \{e'_\alpha\})$ are related by

$$t' = t + \text{const.}$$
$$x'^\alpha = A^\alpha{}_\beta(t)x^\beta + b^\alpha(t)$$

where $A^\alpha{}_\beta(t)$ is an orthogonal matrix (in general time-dependent) and $b^\alpha(t)$ represents the vector from p_0 to p'_0.

(ii) To study dynamics, we need a time derivative for vector fields. The only option is $\pounds_T$ with $T = \frac{\partial}{\partial t}$, as will be seen in section 16.5.

(iii) If $\vec{a}$ is a space-like vector and $\frac{d}{dt} = \pounds_T$ in a Galilean system F, then in a second Galilean system F' we have

$$\frac{d\vec{a}}{dt'} = \frac{d\vec{a}}{dt} + \vec{\omega} \times \vec{a}$$

where $\vec{\omega}$ is interpreted as the 'angular velocity' of F' with respect to F.

16.3 Dynamics of a perfect fluid

Consider a perfect fluid (that is, a non-viscous fluid with adiabatic flow, meaning that it is a thermal insulator) with mass density ρ and velocity $V = (V^x,\ V^y,\ V^z) \in \mathbb{E}^3$. Let $\omega = dx \wedge dy \wedge dz$ be the volume element in $\mathbb{E}^3$.

The motion of the fluid along its flow trajectory suggests the use of the Lie derivative. In fact, if $u = (1,\ V^x,\ V^y,\ V^z)$ is the vector tangent to the flow lines in coordinates (t, x, y, z), then the time derivative along a fluid element's trajectory is $\pounds_u$.

Given $X = X^\alpha e_\alpha \in \mathbb{E}^3$ (with $\alpha \in \{x, y, z\}$),

$$(\pounds_u X)^i = [u, X]^i$$
$$= u^k X^i{}_{,k} - X^k u^i{}_{,k}$$
$$\overset{X^t=0}{=} u^t X^i{}_{,t} + u^\alpha X^i{}_{,\alpha} - X^\alpha u^i{}_{,\alpha}$$
$$\overset{u^t=1}{=} X^i{}_{,t} + u^\alpha X^i{}_{,\alpha} - X^\alpha u^i{}_{,\alpha}$$
$$= \left(\frac{\partial}{\partial t} X\right)^i + (\pounds_V X)^i$$

that is,

$$\pounds_u X = \left(\frac{\partial}{\partial t} + \pounds_V\right) X \tag{16.1}$$

This expression can be generalised for any tensor in $\mathbb{E}^3$ (that is, any tensor without a t component).

Note that

$$\pounds_V(\rho\omega) = d[\rho\omega(V)] + d(\rho\omega)(V)$$
$$= d[\rho\omega(V)]$$
$$= d[\omega(\rho V)] \tag{16.2}$$
$$= \operatorname{div}(\rho V)\omega$$

where the first equality follows from theorem 11.14, the second from the fact that $d\omega = 0$, the third from the fluid's homogeneity, and the fourth from remark 8.6 (iv). Then the continuity equation (conservation of matter)

$$\frac{\partial\rho}{\partial t} + \operatorname{div}(\rho V) = 0 \tag{16.3}$$

is equivalent to $(\frac{\partial}{\partial t} + \pounds_V)(\rho\omega) = 0$ and can be written simply as

$$\pounds_u(\rho\omega) = 0 \tag{16.4}$$

(the drag of a mass element of the fluid along its trajectory does not undergo any changes; this idea is far more intuitive than the derivative from vector calculus).

16.4 Inertial systems and the Newtonian connection

(i) There are certain special systems of reference, called 'inertial', in which the acceleration $\ddot{\vec{x}}$ of a particle with mass m is related to a 'force' vector $\vec{f}$ by

$$m\ddot{\vec{x}} = \vec{f}$$

This relationship is also true in another system of reference F' only if

$$\ddot{A}^\alpha{}_\beta = \dot{A}^\alpha{}_\beta = 0$$
$$\ddot{b}^\alpha = 0$$

that is, this is the transformation law of inertial systems.

(ii) There exists a kind of motion, called 'free fall', characterised by the equation of motion

$$\ddot{\vec{x}} = -\nabla\varphi$$

where φ is independent of the system of reference and is called the 'gravitational' or 'Newtonian potential'. The systems in which the previous equation preserves its form are called 'Newtonian systems of reference' and are related by

$$\ddot{A}^{\alpha}{}_{\beta} = \dot{A}^{\alpha}{}_{\beta} = 0$$

$$\ddot{b}^{\alpha} = 0$$

The free-fall equation can be written as

$$\ddot{x}^{a} + {}^{N}\Gamma^{a}{}_{bc}\dot{x}^{b}\dot{x}^{c} = 0$$

where ${}^{N}\Gamma^{a}{}_{00} = \varphi^{,a}$ (for $\alpha \in \{1, 2, 3\}$) and ${}^{N}\Gamma^{a}{}_{bc} = 0$ in all other cases. ${}^{N}\Gamma^{a}{}_{bc}$ is the Newtonian connection from example 13.8.

16.5 Matter and the curvature of space–time

(i) Let us calculate the Riemann tensor from equation (14.4) for the Newtonian connection:

$$ {}^{N}R^{\alpha}{}_{0\beta0} = \varphi_{,\alpha\beta}$$

$$ {}^{N}R^{\alpha}{}_{00\beta} = -\varphi_{,\alpha\beta}$$

$$ {}^{N}R^{a}{}_{bcs} = 0 \quad \text{in all other cases}$$

Note that ${}^{N}R_{0\alpha\beta0} \neq -{}^{N}R_{\alpha0\beta0}$, so a metric tensor for which ${}^{N}\Gamma$ is the Levi-Civita connection does not exist.

(ii) There is experimental evidence that the Newtonian potential φ satisfies

$$\nabla^{2}\varphi = 4\pi G\rho$$

where $G = \text{const.}$ and ρ characterises the source of φ. From the Riemann tensor above we can calculate the Ricci tensor:

$$ {}^{N}R_{00} = \nabla^{2}\varphi$$

that is,

$$ {}^{N}R_{00} = 4\pi G\rho$$

concluding that matter produces curvature of space–time.

16.6 Hamiltonian dynamics and symplectic manifolds

We saw in section (8.14) that the cotangent bundle $\mathfrak{X}^{*}(M)$ is the natural domain for the study of analytical mechanics on a manifold. We shall see here that a 'symplectic' structure arises naturally in this space and we will construct it.

Let M be the phase space of a physical system and let $H : M \to \mathbb{R}$ be its associated Hamiltonian function. This measures the total energy of the system and is constructed from the kinetic and potential energy contributions

$$H(q, p) = \frac{1}{2m}|p|^2 + V(q)$$

respectively, where $q = (q^1, ..., q^n)$ denotes generalised coordinates, $p = (p_1, ..., p_n)$ denotes the associated conjugate momenta, and m the mass of the system. They are related by

$$p_i(t) = m\,\dot{q}^k(t)\delta_{ik}$$

where $\dot{q}$ denotes its time derivative. Newton's equations also tell us that

$$\dot{p}(t) = -\nabla\, V(q(t))$$

From these equations the dynamics may be written as

$$\dot{q}^i(t) = \delta^{ik}\frac{1}{m}p_k(t) = \frac{\partial H}{\partial p_i}, \qquad \dot{p}_i(t) = -\delta^k{}_i\frac{\partial V}{\partial q^k}(q(t)) = -\frac{\partial H}{\partial q^i} \qquad (16.5)$$

which are the so-called 'Hamilton's equations' for the system.

From H we can define a vector field $X_H \in \mathfrak{X}(M)$:

$$X_H := \left(\frac{\partial H}{\partial p_1}, ...\frac{\partial H}{\partial p_n}, -\frac{\partial H}{\partial q^1}, ...-\frac{\partial H}{\partial q^n}\right)$$

whereby Hamilton's equations take the compact form $(\dot{q}(t), \dot{p}(t)) = X_H(q(t), p(t))$. Note that

$$£_{X_H}H = X_H[H] = g_i{}^k\frac{\partial H}{\partial p_i}\frac{\partial H}{\partial q^k} - g^i{}_k\frac{\partial H}{\partial q^i}\frac{\partial H}{\partial p_k} = 0 \qquad (16.6)$$

that is, the derivative of H in the direction of X_H vanishes, or, equivalently, X_H points in the direction of constant energy.

This is equivalent to saying that

$$dH[X_H] = 0 \qquad (16.7)$$

We could then define a rank $\binom{0}{2}$ tensor $\omega : \mathfrak{X}(M) \times \mathfrak{X}(M) \to \mathbb{R}$ such that $\omega[X_H] = dH \in \mathfrak{X}^*(M)$. Then, for all $X \in \mathfrak{X}(M)$ we have $\omega[X_H, X] = dH[X]$ and, in particular,

$$\omega[X_H, X_H] = 0 \qquad (16.8)$$

But every $X \in \mathfrak{X}(M)$ may be seen *locally* as a Hamiltonian vector field, i.e. may be seen at a given point $p \in M$ as X_H for some Hamiltonian function H, so that ω so defined is really a 2-form in M: $\omega \in \Lambda^2(M)$. In other words, the existence of a non-degenerate 2-form on M, such as ω, corresponds to a dynamics that conserves energy, given by Hamilton's equations. One extra requirement on ω is that

$$\pounds_{X_H}\omega = 0$$

so that it is carried along the system's flow.

(M, ω) is called a 'symplectic manifold', and the phase space manifold M endowed with the structure given to it by ω is the natural scenario for the study of analytical mechanics.

16.7 Thermodynamics

For a single-component ideal gas, the equation of conservation of energy may be written as

$$dU = \delta Q - P\, dV \tag{16.9}$$

where U is the system's internal energy, δQ is the heat it has absorbed, and $P\, dV$ is the work done by it (P is the pressure, V is its volume).

Since dU and dV are 1-forms, δQ must also be a 1-form. Is it an exact 1-form? In other words, does there exist a function $Q = Q(V, U)$ such that $\delta Q = dQ$?

Recall that a thermodynamic system can be characterised by two of its variables, that is, it 'lives' on a two-dimensional manifold M. For the moment, let us take U and V to be these variables.

Suppose there exists a function $Q = Q(U, V) : M \longrightarrow \mathbb{R}$ such that $\delta Q = dQ$. Then $d(dQ) = d^2Q = 0$ and, from equation (16.9) and using $d^2U = 0$ and $dV \wedge dV = 0$, we have

$$0 = d^2Q = d(dQ) = \frac{\partial P}{\partial V}dV \wedge dV + \frac{\partial P}{\partial U}dU \wedge dV = \frac{\partial P}{\partial U}dU \wedge dV$$

that is, $\frac{\partial P}{\partial U} = 0$, which would be very strange for an ideal gas. Therefore, δQ cannot be exact.

Since $\delta Q \in \mathfrak{X}^*(M)$ and it is not exact, there exist two functions

$$T = T(U, V) : M \longrightarrow \mathbb{R}$$
$$S = S(U, V) : M \longrightarrow \mathbb{R}$$

such that $\delta Q = TdS$. (Here we have used Frobenius's theorem; see example 12.12.)

Substituting in equation (16.9)

$$dU = T\, dS - P\, dV \tag{16.10}$$

whereby

$$0 = d^2U = dT \wedge dS - dP \wedge dV \tag{16.11}$$

Taking $T = T(S, V)$ and $P = P(S, V)$ and making use of equation (16.11), we have

$$dT = \frac{\partial T}{\partial V}dV + \frac{\partial T}{\partial S}dS$$

$$dP = \frac{\partial P}{\partial V}dV + \frac{\partial P}{\partial S}dS$$

and thus

$$\frac{\partial T}{\partial V}dV \wedge dS - \frac{\partial P}{\partial S}dS \wedge dV = 0$$

that is,

$$\frac{\partial T}{\partial V} = -\frac{\partial P}{\partial S} \tag{16.12}$$

If we now take $S = S(T, V)$ and $P = P(T, V)$ and again use equation (16.11), we obtain

$$\frac{\partial S}{\partial V} = \frac{\partial P}{\partial T} \tag{16.13}$$

These are two of Maxwell's relations. They can all be trivially obtained from equation (16.11).

In equation (16.12), our variables are S and V, so $\frac{\partial T}{\partial V}$ is calculated at constant S and $\frac{\partial P}{\partial S}$ is calculated at constant V. In equation (16.13), $\frac{\partial S}{\partial V}$ is calculated at constant T and $\frac{\partial P}{\partial T}$ is calculated at constant V.

Let us now take $T = T(P, S)$, $S = S(T, P)$, and $P = P(T, S)$, and denote with a subscript, as is usual, the variable that is held constant in each derivation. Then

$$dT \wedge dS = \left(\frac{\partial T}{\partial P}\right)_S dP \wedge dS + \left(\frac{\partial T}{\partial S}\right)_P dS \wedge dS$$

$$= \left(\frac{\partial T}{\partial P}\right)_S dP \wedge dS$$

$$= -\left(\frac{\partial T}{\partial P}\right)_S dS \wedge dP$$

$$= -\left(\frac{\partial T}{\partial P}\right)_S\left(\frac{\partial S}{\partial T}\right)_P dT \wedge dP$$

$$= \left(\frac{\partial T}{\partial P}\right)_S\left(\frac{\partial S}{\partial T}\right)_P dP \wedge dT$$

$$= \left(\frac{\partial T}{\partial P}\right)_S\left(\frac{\partial S}{\partial T}\right)_P\left(\frac{\partial P}{\partial S}\right)_T dS \wedge dT$$

$$= -\left(\frac{\partial T}{\partial P}\right)_S\left(\frac{\partial P}{\partial S}\right)_T\left(\frac{\partial S}{\partial T}\right)_P dT \wedge dS$$

that is,

$$\left(\frac{\partial T}{\partial P}\right)_S\left(\frac{\partial P}{\partial S}\right)_T\left(\frac{\partial S}{\partial T}\right)_P = -1 \tag{16.14}$$

IOP Publishing

Differential Topology and Geometry with Applications to Physics

Eduardo Nahmad-Achar

Chapter 17

Special relativity, electrodynamics, and the Poincaré group

17.1 The postulates of special relativity

(i) Space–time is a four-dimensional class-C^∞ Lorentzian manifold (M, η).

(ii) There is a family of coordinate systems, called 'inertial', in which the metric has the form $\eta_{ab} = \mathrm{diag}(-1, 1, 1, 1)$ (therefore, $\Gamma^a{}_{bc}$, the Riemann tensor, and the Ricci tensor and scalar are all zero). Any two inertial systems are related by a transformation of the form

$$x'^a = L^a{}_b x^b + d^a$$

where $L^a{}_b = \mathrm{const.}, d^a = \mathrm{const.}$, and $L^T \eta L = \eta$. These transformations form a group, the 'Poincaré group'.

(iii) The light cones of η_{ab} determine the causal structure of M and at the same time describe the propagation of rays of light in vacuum. The trajectories of free particles are time-like geodesics, and along these trajectories

$$\int d\tau = \int \sqrt{-\eta_{ab}\, dx^a\, dx^b} = \int \sqrt{1 - v^2}\, dt$$

is the 'proper time' of the particle in question.

17.2 The electromagnetic tensor and Maxwell's equations

Let $\{x^i\}$ be an inertial coordinate system. If E^α and B^α are, respectively, the components of the electric and magnetic fields, we can define a rank-$\binom{0}{2}$ tensor, the 'Maxwell tensor', as

doi:10.1088/2053-2563/aadf65ch17

$$F_{a0} = E_a$$
$$F_{\alpha\beta} = \varepsilon_{\alpha\beta\gamma} B^\gamma$$
$$F_{ab} = F_{[ab]}$$

where $\varepsilon_{\alpha\beta\gamma}$ is the completely antisymmetric unit 'tensor', commonly called the 'Levi-Civita tensor':

$$\varepsilon_{\alpha\beta\gamma} = \begin{cases} 1 & \text{if } (\alpha, \beta, \gamma) \text{ is an even permutation of } (1, 2, 3) \\ -1 & \text{if } (\alpha, \beta, \gamma) \text{ is an odd permutation of } (1, 2, 3) \\ 0 & \text{otherwise} \end{cases}$$

If ρ is the charge density and j is the current density, we can build a current 4-vector J as follows:

$$J^0 = \rho, \quad J^\alpha = j^\alpha$$

Maxwell's equations are then

$$F_{[ab, c]} = 0 \tag{17.1}$$

$$F^{ab}{}_{,b} = 4\pi J^a \tag{17.2}$$

as we will now see.

We have

$$F_{ab} = \begin{pmatrix} 0 & -E_x & -E_y & -E_z \\ E_x & 0 & B_z & -B_y \\ E_y & -B_z & 0 & B_x \\ E_z & B_y & -B_x & 0 \end{pmatrix} \tag{17.3}$$

and

$$\eta_{ab} = \operatorname{diag}(-1, 1, 1, 1)$$

Thus,

$$F^a{}_b = \begin{pmatrix} 0 & E_x & E_y & E_z \\ E_x & 0 & B_z & -B_y \\ E_y & -B_z & 0 & B_x \\ E_z & B_y & -B_x & 0 \end{pmatrix}, \quad F^{ab} = \begin{pmatrix} 0 & E_x & E_y & E_z \\ -E_x & 0 & B_z & -B_y \\ -E_y & -B_z & 0 & B_x \\ -E_z & B_y & -B_x & 0 \end{pmatrix}$$

whereby we may see that:

(i) $F_{[ab,c]} = dF = 0$ is equivalent to $F_{ab,c} + F_{bc,a} + F_{ca,b} = 0$.

(i1) Taking $a = 1$, $b = 2$ and $c = 3$,

$$0 = B_{z,z} + B_{x,x} + B_{y,y} = \nabla \cdot B \tag{17.4}$$

(i2) Taking $a = 0$, we have the following cases:

(i2.1) $b = 2$, $c = 3$: $\quad -E_{y,z} + B_{x,0} + E_{z,y} = 0$

(i2.2) $b = 3$, $c = 1$: $\quad -E_{z,x} + B_{y,0} + E_{x,z} = 0$

(i2.3) $b = 1$, $c = 2$: $\quad -E_{x,y} + B_{z,0} + E_{y,x} = 0$,

which results in

$$\frac{\partial B}{\partial t} + \nabla \times E = 0 \tag{17.5}$$

(ii) We have $J = (\rho, J^1, J^2, J^3) = (\rho, J)$ and $F^{ab}{}_{,b} = 4\pi J^a$, whereby,

(ii1) Taking $a = 0$, we have

$$4\pi\rho = E_{x,x} + E_{y,y} + E_{z,z} = \nabla \cdot E \tag{17.6}$$

(ii2) Taking $a \neq 0$, we have the following cases:

(ii2.1) $a = 1$: $\quad 4\pi J^1 = -E_{c,0} + B_{z,y} - B_{y,z}$

(ii2.2) $a = 2$: $\quad 4\pi J^2 = -E_{y,0} - B_{z,x} + B_{x,z}$

(ii2.3) $a = 3$: $\quad 4\pi J^3 = -E_{z,0} + B_{y,x} - B_{x,y}$

which results in

$$4\pi J = -\frac{\partial E}{\partial t} + \nabla \times B \tag{17.7}$$

Equations (17.4)–(17.7) are Maxwell's equations in their traditional differential form.

Note that equation (17.1) tells us, through Stokes' theorem, that the integral of F over the boundary of a compact volume is zero, as it is a closed form, while equation (17.2) can be taken as the definition of the current density and the charge density.

We define the 'electromagnetic energy–momentum tensor' T^{ab} as

$$T^{ab} \overset{\text{def}}{=} F^{ac}F^b{}_c - \frac{1}{4}\eta^{ab}F_{cd}F^{cd} \tag{17.8}$$

whereby the Poynting vector S and the energy density W are given by

$$W = \frac{1}{2}(E^2 + B^2) = T^{00} \tag{17.9}$$

$$S = E \times B = T^{0\alpha} \tag{17.10}$$

In the absence of charges and currents, the conservation of energy is written as

$$\dot{W} + \nabla \cdot S = 0$$

that is,

$$T^{0a}{}_{,a} = 0$$

Similarly, the conservation of momentum is written as

$$\dot{S} + T^{\alpha\beta}{}_{,\beta} = 0$$

that is,

$$T^{\alpha\beta}{}_{,\beta} = 0$$

Both conservation laws are included in the equation

$$T^{ab}{}_{,b} = 0 \tag{17.11}$$

17.3 Continuous media

Consider a perfect fluid with energy density ρ, pressure p, and 4-velocity u (assume there is no viscosity). The 'energy–momentum tensor' of the perfect fluid is defined as

$$T^{ab} \stackrel{\text{def}}{=} (\rho + p)u^a u^b + p\eta^{ab} \tag{17.12}$$

By analogy with the electromagnetic field, we would expect $T^{ab}{}_{,b} = 0$ to be a conservation law. What is $T^{ab}{}_{,b}$ for fluids? Let us calculate it:

$$0 = T^{ab}{}_{,b} = (\rho u^b)_{,b} u^a + \rho u^a{}_{,b} u^b + p^{,a} + p_{,b} u^b u^a + p(u^a u^b{}_{,b} + u^a{}_{,b} u^b) \tag{17.13}$$

Now, $\eta^{ij} u_i u_j = u^i u_i = -1$, whereby

$$0 = \left(\eta^{ij} u_i u_j\right)_{,k} u^k$$
$$= \eta^{ij} u_{i,k} u_j u^k + \eta^{ij} u_i u_{j,k} u^k$$
$$= 2\eta^{ij} u_{i,k} u_j u^k$$

and thus $u_{i,k} u^i u^k = 0$. Using this result and multiplying equation (17.13) by u_a, we obtain

$$0 = -(\rho u^b)_{,b} + p^{,a} u_a - p_{,b} u^b - p u^b{}_{,b}$$

that is,

$$0 = \rho_{,a} u^a + (\rho + p)u^a{}_{,a}$$

which can be written as

$$\dot{\rho} + (\rho + p)\theta = 0$$

where $\dot{\rho} \stackrel{\text{def}}{=} \rho_{,a} u^a$ is the 'convective derivative' of ρ along u, and $\theta \stackrel{\text{def}}{=} u^a{}_{,a}$ measures the divergence of u: if ΔV is a volume element of the fluid we have

$$\theta = \frac{(\Delta V)^{\cdot}}{\Delta V}$$

Then

$$\dot{\rho}\Delta V + (\rho + p)(\Delta V)^{\cdot} = 0$$

and, defining $\Delta E = \rho\Delta V$ (the energy contained in a volume element ΔV), $u_a T^{ab}{}_{,b} = 0$ is equivalent to

$$(\Delta E)^{\cdot} + p(\Delta V)^{\cdot} = 0 \qquad (17.14)$$

which is the first law of thermodynamics.

We define $h^{ab} = \eta^{ab} + u^a u^b$ and note that

$$h^{ab} = h^{ba}, \quad h^a{}_a = 3, \quad h^{ab}u_b = 0, \quad h^a{}_b h^b{}_c = h^a{}_c$$

so h^{ab} is a tensor which projects an arbitrary vector onto the subspace perpendicular to u. Projecting equation (17.13) onto this subspace, we have

$$h^{ab} T_b{}^c{}_{,c} = 0$$

that is,

$$(\rho + p)\dot{u}^a + h^{ab}p_{,b} = 0 \qquad (17.15)$$

which is Euler's equation for a perfect fluid with mass density $\rho + p$.

17.4 More on the flow of perfect fluids

In section 16.7 we wrote the conservation of energy for a thermodynamic system (called the 'first law of thermodynamics') as

$$dU = \delta Q - p\, dV$$

with $\delta Q = TdS$, U being the internal energy of the system, V its volume, and p, T, S the pressure, temperature, and entropy in the volume V, respectively.

In special relativity (and the results here will continue to hold in general relativity) the mass is different for different observers so that, thinking of a perfect fluid, we should write $d(\rho V)$ in the place of dU, where ρ is the mass density of the fluid. Then

$$d(\rho V) = -p\, dV + T\, dS$$

What *is* conserved is the total number of particles N of the fluid, thinking of it microscopically, since it is a scalar quantity. Writing $n = N/V$ we have, in terms of n (and dividing by N),

$$d(\rho/n) = -p\, d\left(\frac{1}{n}\right) + T\, d\left(\frac{S}{N}\right) = -p\, d\left(\frac{1}{n}\right) + T\, ds$$

where $s = S/N$ is the entropy per particle. Then

$$\frac{d\rho}{n} - \frac{\rho}{n^2}\,dn = p\,\frac{1}{n^2}\,dn + T\,ds$$

i.e.

$$d\rho = \frac{p + \rho}{n}dn + n\,T\,ds \qquad (17.16)$$

On the other hand, the energy–momentum tensor for a perfect fluid is (see section 17.3)

$$T^{ab} = (p + \rho)u^a u^b + p\,\eta^{ab}$$

and satisfies $T^{ab}{}_{,b} = 0$. In the fluid's reference frame we then have

$$
\begin{aligned}
0 &= u_a T^{ab}{}_{,b} \\
&= T^{0b}{}_{,b} \\
&= (p + \rho)_{,b}u^b u^0 + (p + \rho)[u^b{}_{,b}u^0 + u^0{}_{,b}u^b] + p_{,b}\eta^{0b} + p\,\eta^{0b}{}_{,b} \\
&= \frac{d}{dt}(p + \rho) + (p + \rho)u^b{}_{,b} - \frac{d}{dt}p \\
&= \frac{d\rho}{dt} + (p + \rho)u^b{}_{,b}
\end{aligned}
\qquad (17.17)
$$

where we have used $u^0{}_{,b} = u_a u^a{}_{,b} = (1/2)(u_a u^a)_{,b} = 0$.

In order to calculate $u^b{}_{,b}$ we note that

$$(n\,u^a)_{,a} = n_{,a}u^a + n\,u^a{}_{,a} = \frac{dn}{dt} + n\,u^a{}_{,a}$$

which vanishes since the particle's flux density nu^a is conserved in the fluid's reference frame. Therefore

$$u^a{}_{,a} = -\frac{1}{n}\frac{dn}{dt}$$

and substituting into equation (17.17)

$$\frac{d\rho}{dt} = \frac{p + \rho}{n}\frac{dn}{dt} \qquad (17.18)$$

Comparing equations (17.16) and (17.18) we have

$$\frac{ds}{dt} = 0$$

so that the flow of a perfect fluid is isentropic, i.e. the entropy per particle is conserved.

17.5 Energy–momentum tensor

It would seem that in special relativity we may assign to any field or particle system an 'energy–momentum tensor' T_{ab} such that $T^{ab}{}_{,b} = 0$ gives us certain conservation laws.

Indeed, consider a system whose action is of the form

$$S = \int \Lambda\left(q, \frac{\partial q}{\partial x^a}\right) dV \, dt$$

where $\Lambda = \Lambda\left(q, \frac{\partial q}{\partial x^a}\right)$ is a function of the quantities q which describe the state of the system, and of their first derivatives with respect to the space coordinates and time. Note that the Lagrangian of the system is $\int \Lambda \, dV$. Thus, we may call Λ the 'Lagrangian density'.

The field equations (or equations of motion of the system) are obtained in the usual way:

$$0 = \delta S$$

$$= \int \left[\frac{\partial \Lambda}{\partial q}\delta q + \frac{\partial \Lambda}{\partial q_{,a}}\delta(q_{,a})\right] dV \, dt$$

$$= \int \left[\frac{\partial \Lambda}{\partial q}\delta q + \frac{\partial}{\partial x^a}\left(\frac{\partial \Lambda}{\partial q_{,a}}\delta q\right) - \delta q \frac{\partial}{\partial x^a}\frac{\partial \Lambda}{\partial q_{,a}}\right] dV \, dt$$

The second term of the integrand is a total divergence and hence does not contribute to δS (by virtue of Gauss's theorem). We therefore have

$$\frac{\partial}{\partial x^a}\frac{\partial \Lambda}{\partial q_{,a}} - \frac{\partial \Lambda}{\partial q} = 0 \tag{17.19}$$

If we now write

$$\frac{\partial \Lambda}{\partial x^a} = \frac{\partial \Lambda}{\partial q}\frac{\partial q}{\partial x^a} + \frac{\partial \Lambda}{\partial q_{,b}}\frac{\partial q_{,b}}{\partial x^a}$$

and substitute this in the equations of motion obtained above, we obtain

$$\delta^b_a\frac{\partial \Lambda}{\partial x^b} = \frac{\partial \Lambda}{\partial x^a} = \frac{\partial}{\partial x^b}\left(\frac{\partial \Lambda}{\partial q_{,b}}\right)q_{,a} + \frac{\partial \Lambda}{\partial q_{,b}}\frac{\partial q_{,b}}{\partial x^a} = \frac{\partial}{\partial x^b}\left(q_{,a}\frac{\partial \Lambda}{\partial q_{,b}}\right)$$

By defining

$$T_a{}^b = q_{,a}\frac{\partial \Lambda}{\partial q_{,b}} - \delta^b_a\Lambda \tag{17.20}$$

the previous equation is reduced to

$$T^b_{a,\,b} = 0 \tag{17.21}$$

The tensor $T^{ab} = \eta^{ac}T_c{}^b$ is the so-called 'energy–momentum tensor' of the system, and the last equation gives us the conservation of energy of the system. Note that the way to define T^{ab} is not unique, for, by virtue of Gauss's theorem, if we add to it the divergence of any antisymmetric tensor, the relationship $T^b_{a,\,b} = 0$ will hold and the system will have the same energy and momentum. Note also that T^{ab} is not, in general symmetric, although it does differ from a symmetric tensor by the total divergence of an antisymmetric tensor. It is therefore possible to obtain a symmetric tensor directly, as will be seen later on (see remark 18.6).

IOP Publishing

Differential Topology and Geometry with Applications to Physics

Eduardo Nahmad-Achar

Chapter 18

General relativity

In this chapter we will derive the equations for the gravitational field, that is, the basic equations of general relativity, also known as Einstein's equations. These equations, as we shall see, are nonlinear, so the superposition principle is not valid. They implicitly contain the equations governing the matter which produces the gravitational field, so the distribution of matter is not arbitrary, but rather determined by the field equations themselves.

18.1 Equivalence principle

General relativity arises as an extension of special relativity to accelerated systems of reference. Galileo's experiments show that an object's acceleration due to an external gravitational field is independent of the object itself.

This allows us to identify accelerated systems of reference with gravitational fields: an accelerated system is equivalent to an inertial system in the presence of a gravitational field. Thus, the concept of 'inertial system' loses its meaning and general relativity becomes a theory of gravitation.

18.2 Principle of relativity

In order to be able to consider any system of reference as equivalent for studying the laws of physics, it is necessary to extend the group of Lorentz transformations to the group of continuous coordinate transformations (both linear and nonlinear).

In other words, the equations expressing the laws of nature must be covariant (that is, they must maintain their form) under continuous coordinate transformations. This is the so-called 'principle of relativity'.

Under the group of all continuous coordinate transformations, the only invariant thing is that nearby points have similar coordinates.

Suppose we have two points p, q in the Minkowski space–time (with Cartesian coordinates) which are infinitesimally close to each other. To express the relationship between p and q, special relativity tells us that the speed of light c is constant and that the distance ds between p and q is

$$\sqrt{(dx^1)^2 + (dx^2)^2 + (dx^3)^2 - c^2(dx^0)^2} = ds$$

that is,

$$-c^2(dx^0)^2 + (dx^1)^2 + (dx^2)^2 + (dx^3)^2 = ds^2$$

or

$$\eta_{ab}dx^a dx^b = ds^2 \tag{18.1}$$

where $\eta_{ab} = \text{diag}(-1, 1, 1, 1)$ is the metric of the space of special relativity and we have set $c = 1$. In general, for an arbitrary Lorentzian manifold we have

$$g_{ab}dx^a dx^b = ds^2 \tag{18.2}$$

where g_{ab} is the manifold's metric tensor.

Equation (18.2) is a tensor equation and thus holds in any coordinate system. In other words, the components g_{ab} of the metric tensor determine the geometric properties of each coordinate system (whether curvilinear or not), and from the identification of systems of reference with gravitational fields (given by the equivalence principle; see remark 18.1) we have that the quantities g_{ab} determine the gravitational field.

18.3 Action for the gravitational field

If the metric g_{ab} determines the gravitational field, it is important to ask what the dynamics of g_{ab} are, that is, what the field equations are.

To obtain the gravitational field equations, we must first obtain the action of the field itself, S_g. The total action S will be

$$S = S_g + S_m$$

where S_m is the action of all 'matter fields' (including the electromagnetic field). Here, anything other than the gravitational field will be called a 'matter field'.

Normally, the Lagrangian depends on the field variables and their first derivatives, which in our case would be g_{ab} and $\Gamma^a{}_{bc}$. However, it is immediately obvious that it is impossible to build an invariant scalar from g_{ab} and $\Gamma^a{}_{bc}$, since an appropriate choice of coordinates will render $\Gamma^a{}_{bc}$ zero at an arbitrary point (see theorem 13.6).

We must therefore include the second derivatives of g_{ab}. It is convenient to consider a Lagrangian $L = L(g_{ab}; g_{ab,c}; g_{ab;cd})$ which is linear in the second derivatives of g_{ab}, as we will then, by virtue of Gauss's theorem, be able to add a total divergence and possibly transform the action integral into an expression without second derivatives, resulting in second-order field equations. The only invariant we can build from g_{ab} and its first and second derivatives which is linear in the latter is the Ricci scalar R.

We thus arrive at

$$S_g = -\frac{1}{16\pi} \int R\sqrt{-g}\,d\Omega \tag{18.3}$$

where $d\Omega$ is the volume element of the manifold, $g = \det(g_{ab})$, and the constant $-\frac{1}{16\pi}$ is chosen in order to recover the Newtonian limit in the weak-field approximation.

18.4 Lemma

Let $g = \det(g_{ab})$, and let A^i be an arbitrary vector field. Then

$$dg = gg^{ik}dg_{ik} = -gg_{ik}dg^{ik} \qquad (18.4)$$

$$\Gamma^i{}_{ki} = \frac{\partial}{\partial x^k}\ln\sqrt{-g} \qquad (18.5)$$

$$A^i{}_{;i} = \frac{1}{\sqrt{-g}}\frac{\partial}{\partial x^i}\left(\sqrt{-g}\,A^i\right) \qquad (18.6)$$

Proof. We can calculate dg by taking the differential of each component of (g_{ab}) and multiplying it by its coefficient in the expression for the determinant of (g_{ab}), that is, for the corresponding minor. However, the components of (g^{ab}) are precisely those minors divided by the determinant itself; in other words, the minors of (g_{ab}) are gg^{ik}. Therefore,

$$dg = gg^{ik}dg_{ik}$$

Now, $g_{ik}g^{ik} = \delta^i{}_i = 4$, from where it follows immediately, taking the differential, that $g^{ik}dg_{ik} = -g_{ik}dg^{ik}$. Therefore,

$$dg = -gg_{ik}dg^{ik}$$

Now,

$$\Gamma^i{}_{ki} = \frac{1}{2}g^{im}(g_{km,\,i} + g_{mi,\,k} - g_{ki,\,m})$$

$$= \frac{1}{2}g^{im}g_{mi,\,k}$$

whereby, using equation (18.4),

$$\Gamma^i{}_{ki} = \frac{1}{2g}\frac{\partial g}{\partial x^k} = \frac{\partial}{\partial x^k}\ln\sqrt{-g}$$

Finally, using equation (18.5),

$$A^i{}_{;i} = \frac{\partial A^i}{\partial x^i} + \Gamma^i{}_{ki}A^k = \frac{\partial A^i}{\partial x^i} + A^k\frac{\partial}{\partial x^k}\ln\sqrt{-g} = \frac{1}{\sqrt{-g}}\frac{\partial}{\partial x^i}\left(A^i\sqrt{-g}\right)$$

$\square$

18.5 Einstein's equations in a vacuum

Suppose there are no external fields (whether electromagnetic fields or matter fields), that is, $S_m = 0$. The gravitational field equations in vacuum can then be obtained from

$$\delta S_g = 0 \tag{18.7}$$

Calculating the variation,

$$
\begin{aligned}
0 &= -\frac{q}{16\pi}\delta \int R\sqrt{-g}\, d\Omega \\
&= -\frac{1}{16\pi}\delta \int g^{ik}R_{ik}\sqrt{-g}\, d\Omega \\
&= -\frac{1}{16\pi}\int \left(R_{ik}\sqrt{-g}\,\delta g^{ik} + R_{ik}g^{ik}\delta\sqrt{-g} + g^{ik}\sqrt{-g}\,\delta R_{ik} \right) d\Omega
\end{aligned}
\tag{18.8}
$$

For the second term, we use equation (18.4):

$$\delta\sqrt{-g} = \frac{1}{2\sqrt{-g}}\delta g = -\frac{1}{2}\sqrt{-g}\,g_{ik}\delta g^{ik} \tag{18.9}$$

For the third term, note that, using normal coordinates, $\Gamma^i{}_{jk} = 0$ (see theorem 13.6) at an arbitrary point. Therefore,

$$
\begin{aligned}
R_{ik} &= \Gamma^l{}_{ik,\,l} - \Gamma^l{}_{il,\,k} \\
\delta R_{ik} &= \delta\Gamma^l{}_{ik,\,l} - \delta\Gamma^l{}_{il,\,k} \\
g^{ik}\delta R_{ik} &= g^{ik}\left(\delta\Gamma^l{}_{ik,\,l} - \delta\Gamma^l{}_{il,\,k}\right) \\
&= g^{ik}\frac{\partial}{\partial x^l}\delta\Gamma^l{}_{ik} - g^{il}\frac{\partial}{\partial x^l}\delta\Gamma^k{}_{ik} \\
&= \frac{\partial}{\partial x^l}\omega^l \\
&\overset{\text{eq. (18.6)}}{=} \frac{1}{\sqrt{-g}}\frac{\partial}{\partial x^l}\left(\sqrt{-g}\,\omega^l\right)
\end{aligned}
$$

where we have defined $\omega^l \overset{\text{def}}{=} g^{ik}\delta\Gamma^l{}_{ik} - g^{il}\delta\Gamma^k{}_{ik}$ and noted that

$$0 = g_{ik;l} = \frac{\partial g_{ik}}{\partial x^l} - g_{mk}\Gamma^m{}_{il} - g_{im}\Gamma^m{}_{kl} = \frac{\partial g_{ik}}{\partial x^l}$$

(because we are calculating these quantities in normal coordinates), whereby the terms $\frac{\partial}{\partial x^l}g^{ik}$ do not contribute to the variation of the action. As ω^l is a vector ($\Gamma^i{}_{jk}$ is not a tensor but $\delta\Gamma^i{}_{jk}$ is (see lemma 12.8)), to transform $\omega^l{}_{,\,l} = 0$ to any coordinate system we write $\omega^l{}_{;\,l} = 0$. In other words,

$$\sqrt{-g}\,g^{ik}\delta R_{ik} = \frac{\partial}{\partial x^l}\left(\sqrt{-g}\,\omega^l\right) \tag{18.10}$$

which is a total divergence and therefore does not contribute to the variation. Substituting equation (18.9) in equation (18.8),

$$0 = \delta S_g = -\frac{1}{16\pi}\int\left(R_{ik} - \frac{1}{2}Rg_{ik}\right)\sqrt{-g}\,\delta g^{ik}\,d\Omega \tag{18.11}$$

whereby the gravitational field equations, called 'Einstein's equations', in a vacuum are

$$R_{ik} - \frac{1}{2}Rg_{ik} = 0 \tag{18.12}$$

18.6 The energy–momentum tensor

In the presence of matter fields (including electromagnetic fields), we need the action S_m for the field itself. In section 17.5 we saw how to calculate the energy–momentum tensor of any physical system determined by an action integral. In general, in curvilinear coordinates the action integral is written as

$$S_m = \int \Lambda\sqrt{-g}\,d\Omega \tag{18.13}$$

Under a coordinate transformation

$$x'^i = x^i + \xi^i$$

where the quantities ξ^i are infinitesimal, we have

$$g'^{ik}(x'') = g^{lm}(x^l)\frac{\partial x'^i}{\partial x^l}\frac{\partial x'^k}{\partial x^m}$$

$$= g^{lm}(x^l)\left(\delta_l^i + \frac{\partial \xi^i}{\partial x^l}\right)\left(\delta_m^k + \frac{\partial \xi^k}{\partial x^m}\right)$$

$$= g^{ik}(x^l) + g^{im}\frac{\partial \xi^k}{\partial x^m} + g^{kl}\frac{\partial \xi^i}{\partial x^l} + \mathcal{O}(\xi^2)$$

On the other hand,

$$g'^{ik}(x'') = g'^{ik}(x^l + \xi^l)$$

$$= g'^{ik}(x^l) + \xi^l\frac{\partial g'^{ik}}{\partial x^l} + \mathcal{O}(\xi^2)$$

$$= g'^{ik}(x^l) + \xi^l\frac{\partial g^{ik}}{\partial x^l} + \mathcal{O}(\xi^2)$$

Therefore,

$$g'^{ik}(x^l) = g^{ik}(x^l) - \xi^l \frac{\partial g^{ik}}{\partial x^l} + g^{il}\frac{\partial \xi^k}{\partial x^l} + g^{kl}\frac{\partial \xi^i}{\partial x^l} \qquad (18.14)$$

The last three terms on the right-hand side of equation (18.14) can be written as $\xi^{k;i} + \xi^{i;k}$. Therefore, we have

$$g'^{ik} = g^{ik} + \delta g^{ik} \qquad (18.15)$$

where $\delta g^{ik} = \xi^{i;k} + \xi^{k;i}$. Asking that $g'_{il}g'^{kl} = \delta_i{}^k$, we also have

$$g'_{ik} = g_{ik} + \delta g_{ik} \qquad (18.16)$$

where $\delta g_{ik} = -\xi_{i;k} - \xi_{k;i}$. Note that, if the transformation $x^i \longrightarrow x'^i = x^i + \xi^i$ is such that $\xi_{i;k} + \xi_{k;i} = 0$, then $\delta g_{ik} = 0$ (i.e. the metric does not change). The equations $\xi_{i;k} + \xi_{k;i} = 0$ are called 'Killing's equations', and the corresponding vector ξ^i is called a 'Killing vector'. Killing's equations are equivalent to

$$\pounds_\xi g_{ab} = 0$$

so they represent symmetry properties of the manifold M. The set of Killing vectors on M forms an algebra over $\mathbb{R}$, and the group of diffeomorphisms generated by these vectors is the corresponding Lie group of isometries of M.

On one hand, we know that $\delta S_m = 0$, as S_m is a scalar. On the other hand, we can write an explicit equation for δS_m: if we denote by q the variables describing the matter field, we have $S_m = S_m(q, g^{ik})$ and the 'equations of motion' of the system are given by

$$\delta S_m/\delta q, \, \delta q' = 0$$

Therefore, given the equations of motion, we do not need to calculate the variation of S_m with respect to q (because said variation is zero), only its variation with respect to g^{ik}. The total variation of S_m is then (see equation (18.13))

$$\delta S_m = \int \left[\frac{\partial \sqrt{-g}\Lambda}{\partial g^{ik}}\delta g^{ik} + \frac{\partial \sqrt{-g}\Lambda}{\partial \left(\dfrac{\partial g^{ik}}{\partial x^l}\right)}\delta\left(\frac{\partial g^{ik}}{\partial x^l}\right) \right] d\Omega$$

$$= \int \left[\frac{\partial \sqrt{-g}\Lambda}{\partial g^{ik}} - \frac{\partial}{\partial x^l}\frac{\partial \sqrt{-g}\Lambda}{\partial \left(\dfrac{\partial g^{ik}}{\partial x^l}\right)} \right] \delta g^{ik} \, d\Omega$$

where we have used Gauss's theorem for the second equality. We define the '(symmetric) energy-momentum tensor' T_{ik} of the system as

$$\frac{1}{2}\sqrt{-g}\,T_{ik} \stackrel{\text{def}}{=} \frac{\partial \sqrt{-g}\,\Lambda}{\partial g^{ik}} - \frac{\partial}{\partial x^l}\frac{\partial \sqrt{-g}\,\Lambda}{\partial\left(\frac{\partial g^{ik}}{\partial x^l}\right)} \tag{18.17}$$

The total variation of S_m is thus

$$\delta S_m = \frac{1}{2}\int T_{ik}\delta g^{ik}\sqrt{-g}\,d\Omega = -\frac{1}{2}\int T^{ik}\delta g_{ik}\sqrt{-g}\,d\Omega \tag{18.18}$$

Recalling that $dg^{ik} = \xi^{i;k} + \xi^{k;i}$ (see equation (18.15)) and using the fact that T_{ik} is symmetric,

$$\begin{aligned}
\delta S_m &= \frac{1}{2}\int T_{ik}(\xi^{i;k} + \xi^{k;i})\sqrt{-g}\,d\Omega \\
&= \int T_{ik}\xi^{i;k}\sqrt{-g}\,d\Omega \\
&= \left(T_i{}^k\xi^i\right)_{;k}\sqrt{-g}\,d\Omega - \int T_i{}^k{}_{;k}\xi^i\sqrt{-g}\,d\Omega \\
&\stackrel{\text{eq. (18.6)}}{=} \int \frac{\partial}{\partial x^k}\left(\sqrt{-g}\,T_i{}^k\xi^i\right)d\Omega - \int T_i{}^k{}_{;k}\xi^i\sqrt{-g}\,d\Omega
\end{aligned}$$

Using Gauss's theorem and the fact that ξ^i is zero at the integration limits,

$$0 = \delta S_m = -\int T_i{}^k{}_{;k}\xi^i\sqrt{-g}\,d\Omega \tag{18.19}$$

whereby, since ξ^i is arbitrary,

$$T^{ik}{}_{;k} = 0 \tag{18.20}$$

This is the covariant form of the conservation law we had already seen in chapter 17 for any metric manifold.

18.7 Einstein's equations

We now have all the necessary elements to derive Einstein's equations in the most general case, that of an arbitrary matter field in the presence of a gravitational field. These equations are obtained from the minimum-action principle, $\delta S = 0$, for the total action $S = S_m + S_g$.

From equation (18.11),

$$\delta S_g = -\frac{1}{16\pi}\int\left(R_{ik} - \frac{1}{2}Rg_{ik}\right)\sqrt{-g}\,\delta g^{ik}\,d\Omega \tag{18.21}$$

From equation (18.18),

$$\delta S_m = \frac{1}{2}\int T_{ik}\delta g^{ik}\sqrt{-g}\,d\Omega \tag{18.22}$$

Therefore, the minimum-action principle $\delta(S_m + S_g) = 0$ gives us

$$R_{ik} - \frac{1}{2}Rg_{ik} = 8\pi T_{ik} \tag{18.23}$$

which are Einstein's equations.

18.8 Example (Schwarzschild metric)

The Schwarzschild metric is a static solution with spherical symmetry to Einstein's equations in a vacuum.

Because of the spherical symmetry, the manifold M is locally of the form $M = M^2 \times \mathbb{S}^2$, where $\mathbb{S}^2$ is the two-dimensional sphere with metric $r^2[d\theta^2 + \sin^2\theta \, d\phi^2]$ and M^2 is a two-dimensional manifold with metric $-B(r) \, dt^2 + A(r) \, dr^2$ ($x^0 = t$, $x^1 = r$, $x^2 = \theta$, $x^3 = \phi$).

A and B do not depend on t because the solution is static, and they do not depend on θ or ϕ because of the spherical symmetry. Because the metric is static, we do not have a $dt \, dr$ term; such a term would change sign under a transformation of the form $t \longrightarrow -t$. Then

$$d\tau^2 = -B(r)dt^2 + A(r)dr^2 + r^2[d\theta^2 + \sin^2\theta \, d\phi^2] \tag{18.24}$$

Using $\Gamma^i_{jk} = \frac{1}{2}g^{im}[g_{mj,k} + g_{mk,j} - g_{jk,m}]$ we calculate the connection coefficients:

$$\Gamma^r_{rr} = \frac{1}{2A}\frac{dA}{dr}$$

$$\Gamma^r_{\phi\phi} = -\frac{r\sin^2\theta}{A}$$

$$\Gamma^r_{tt} = \frac{1}{2A}\frac{dB}{dr}$$

$$\Gamma^r_{\theta\theta} = -\frac{r}{A}$$

$$\Gamma^\theta_{r\theta} = \Gamma^\theta_{\theta r} = \frac{1}{r}$$

$$\Gamma^\theta_{\phi\phi} = -\sin\theta\cos\theta$$

$$\Gamma^\phi_{\theta\phi} = \Gamma^\phi_{\phi\theta} = \cot\theta$$

$$\Gamma^\phi_{\phi r} = \Gamma^\phi_{r\phi} = \frac{1}{r}$$

$$\Gamma^t_{tr} = \Gamma^t_{rt} = \frac{1}{2B}\frac{dB}{dr}$$

and the rest are zero.

From $R_{jk} = \Gamma^i_{ji,k} - \Gamma^i_{jk,i} + \Gamma^m_{ji}\Gamma^i_{km} - \Gamma^m_{jk}\Gamma^i_{im}$, we obtain the elements of the Ricci tensor:

$$R_{rr} = \frac{B''}{2B} - \frac{1}{4}\frac{B'}{B}\left(\frac{A'}{A} + \frac{B'}{B}\right) - \frac{1}{r}\frac{A'}{A}$$

$$R_{\theta\theta} = -1 + \frac{r}{2A}\left(-\frac{A'}{A} + \frac{B'}{B}\right) + \frac{1}{A}$$

$$R_{\phi\phi} = R_{\theta\theta}\sin^2\theta$$

$$R_{tt} = -\frac{B''}{2A} + \frac{1}{4}\frac{B'}{A}\left(\frac{A'}{A} + \frac{B'}{B}\right) - \frac{1}{r}\frac{B'}{A}$$

and the rest are zero. Here, $A' = \frac{dA}{dr}$, $B' = \frac{dB}{dr}$, etc.

Einstein's equations in vacuum are

$$R_{ij} - \frac{1}{2}Rg_{ij} = 0$$

Contracting with g^{ij} and recalling that $g^{ij}g_{ij} = \delta^i_j = 4$, we obtain $R - 2R = 0$, that is, $R = 0$, whereby Einstein's equations become

$$R_{ij} = 0 \tag{18.25}$$

Taking

$$0 = \frac{R_{rr}}{A} + \frac{R_{tt}}{B} = -\frac{1}{rA}\left(\frac{A'}{A} + \frac{B'}{B}\right)$$

we have $\frac{A'}{A} = -\frac{B'}{B}$; to put this another way, $0 = A'B + AB' = (AB)'$, which implies that AB is constant.

The boundary condition for $r \to \infty$ is that the metric approaches the Minkowski metric, so

$$\lim_{r\to\infty} A(r) = \lim_{r\to\infty} B(r) = 1$$

Therefore, $A = \frac{1}{B}$. Inserting this result in the expressions for R_{rr} and $R_{\theta\theta}$ we obtain

$$R_{\theta\theta} = -1 + rB' + B$$

$$R_{rr} = \frac{B''}{2B} + \frac{B'}{rB} = \frac{1}{2rB}R_{\theta\theta}{}'$$

whereby $R_{ij} = 0$ is reduced to $R_{\theta\theta} = 0$:

$$R_{\theta\theta} = 0 \;\Rightarrow\; 1 = rB' + B = (rB)' \;\Rightarrow\; rB = r + \text{const.} \;\Rightarrow\; B = 1 + \frac{\text{const.}}{r}$$

The constant which appears in the above series of implications can be fixed by boundary conditions. Far away from a mass m, we know that $g_{tt} \to -1 - 2\Phi$, where $\Phi = -\frac{mG}{r}$ is the Newtonian potential caused by the mass and G is the universal gravitational constant. Choosing units in which $G = 1$,

$$B = 1 - \frac{2m}{r}$$

Therefore,

$$d\tau^2 = -\left(1 - \frac{2m}{r}\right)dt^2 + \left(1 - \frac{2m}{r}\right)^{-1}dr^2 + r^2[d\theta^2 + \sin^2\theta \, d\phi^2] \qquad (18.26)$$

is the 'Schwarzschild metric'.

Note that this metric shows a singularity at

$$r_s = 2m \qquad (18.27)$$

known as the 'Schwarzschild radius'.

For example, in standard units, the Sun has $r_s \approx 3$ km and a proton has $r_s \approx 10^{-50}$ cm. For all known objects, r_s is within the object, so the Schwarzschild metric is valid from the boundary outwards.

The singularity $r = r_s$ is not, however, a real singularity in space–time M: all the components of the Riemann and Ricci tensors are well defined at $r = 2m$. The singularity is an issue with the coordinate system, and it is possible to eliminate it with a coordinate transformation. For example, if we define

$$\bar{r} \overset{\text{def}}{=} \int \frac{1}{1 - \dfrac{2m}{r}} dr = r + 2m \ln(r - 2m)$$

and

$$v \overset{\text{def}}{=} t + \bar{r}$$

(v is a 'retarded' null coordinate), the Schwarzschild metric becomes

$$d\tau^2 = -\left(1 - \frac{2m}{r}\right)dv^2 + 2\, dv\, dr + r^2[d\theta^2 + \sin^2\theta \, d\phi^2] \qquad (18.28)$$

which is nonsingular for $0 < r < \infty$. We can also define an 'advanced' null coordinate

$$u \overset{\text{def}}{=} t - \bar{r}$$

with which the metric is

$$d\tau^2 = -\left(1 - \frac{2m}{r}\right)dv\, du + r^2[d\theta^2 + \sin^2\theta \, d\phi^2]$$

These are the Eddington–Finkelstein coordinates.

Figure 18.1 shows the curves along which v is constant as solid lines and those along which u is constant as dotted lines in a t versus r diagram. Because u and v are null coordinates, it is easy to visualise the light cones. Note that these are rotated when they cross the surface $r = 2m$: the r coordinate becomes a temporal coordinate for $r < 2m$ (as can be seen in the expression for the metric), so an observer starting at

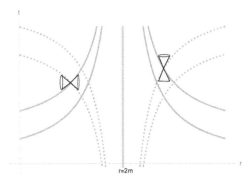

Figure 18.1. Eddington–Finkelstein coordinates for the Schwarzschild metric.

$r > 2m$ and crossing $r = 2m$ can only continue falling towards $r = 0$, never return. The surface $r = 2m$ is then a horizon, the 'Schwarzschild horizon'.

Another interesting way of looking at the Schwarzschild metric, and in particular at the apparent singularity at $r = 2m$, is the following:

Let us take again the hyperplane $\theta = \pi/2$ and follow the trajectory of a light ray with $d\phi = 0$. (This is a null geodesic, as we shall see below in section 18.9.) It is given by

$$
\begin{aligned}
d\tau^2 &= -\left(1 - \frac{2m}{r}\right)dt^2 + \left(1 - \frac{2m}{r}\right)^{-1}dr^2 \\
&= -\left(1 - \frac{2m}{r}\right)\left(dt - \frac{dr}{1 - 2\,m/r}\right)\left(dt + \frac{dr}{1 - 2\,m/r}\right)
\end{aligned}
\tag{18.29}
$$

which motivates the definition of new coordinates $\bar{u}$ and $\bar{v}$ for which

$$
d\bar{u} = dt - \frac{dr}{1 - 2\,m/r}
\tag{18.29a}
$$

$$
d\bar{v} = dt + \frac{dr}{1 - 2\,m/r}
\tag{18.29b}
$$

From these, we have

$$
\frac{1}{2}(d\bar{v} - d\bar{u}) = \frac{dr}{1 - 2\,m/r}
$$

and integrating

$$
\frac{\bar{v} - \bar{u}}{2} = r + 2m \ln\left(\frac{r}{2m} - 1\right)
$$

We now divide by $2m$ and take the exponential

$$
e^{(\bar{v}-\bar{u})/4m} = e^{r/2m}\left(\frac{r}{2m} - 1\right) = e^{r/2m}\left(1 - \frac{2m}{r}\right)\frac{r}{2m}
$$

and substitute into $d\tau^2$, to obtain

$$d\tau^2 = -\frac{2m}{r}e^{-r/2m}e^{(\bar{v}-\bar{u})/4m}du\ dv$$

$$= -\frac{32\ m^3}{r}e^{-r/2m}\left(e^{-\bar{u}/4m}\frac{d\bar{u}}{4m}\right)\left(e^{\bar{v}/4m}\frac{d\bar{v}}{4m}\right) \tag{18.30}$$

$$= -\frac{32\ m^3}{r}e^{-r/2m}d\bar{U}\ d\bar{V}$$

where we have defined

$$\bar{U} = -e^{-\bar{u}/4m}, \qquad \bar{V} = e^{\bar{v}/4m} \tag{18.31}$$

$(\bar{U},\ \bar{V},\ \theta,\ \phi)$ are the so-called 'Kruskal coordinates', or 'Kruskal–Szekeres coordinates'. Note the following:
(a) $\bar{U}$ and $\bar{V}$ are well defined for $r = 2m$, and allow us to extend the Schwarzschild metric for $r < 2m$.
(b) $\bar{U}\bar{V} = e^{r/2m}(1 - r/2m)$, so that $r = 2\ m \Rightarrow \bar{U} = 0$ or $\bar{V} = 0$, and $r = 0 \Rightarrow \bar{U}\bar{V} = 1$. This means that in an extended plot of the Schwarzschild metric, in terms of the coordinates $(\bar{U},\ \bar{V})$, the $\bar{U}$ and $\bar{V}$ axes correspond to the horizon $r = 2m$, while the singularity at $r = 0$ will be given by the hyperbola $\bar{U}\bar{V} = 1$, as shown in figure 18.2.

We draw the $(\bar{U},\ \bar{V})$-coordinates along the diagonals, as they are null coordinates. The original Schwarzschild metric corresponds to the quadrant on the right. Since the horizons $\bar{U} = 0$ and $\bar{V} = 0$ are null, a particle in the top quadrant cannot escape from that region (not even a light ray!); this quadrant is then the interior of a 'black hole'.

The bottom quadrant would represent a 'white hole', from which any particle or light ray eventually escapes. The left quadrant is a copy of the Schwarzschild region, although it cannot communicate with the latter; it represents a universe disconnected from that in the right quadrant.

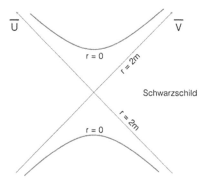

Figure 18.2. Kruskal coordinates for the Schwarzschild metric.

The point $r = 0$ continues to be a singularity, whether as seen in the Scwarzschild coordinates, or in the Eddington–Finkelstein or Kruskal coordinates. To see that it is indeed a real singularity of the manifold M we can calculate the Riemann tensor from the connection coefficients (as we did above for the Ricci tensor). The components of the Riemann tensor that do not vanish are

$$R^t{}_{rtr} = \frac{2m}{r^3} \tag{18.32a}$$

$$R^t{}_{\theta t\theta} = R^t{}_{\phi t\phi} = -\frac{m}{r^3} \tag{18.32b}$$

$$R^\theta{}_{r\theta r} = R^\phi{}_{r\phi r} = -\frac{m}{r^3} \tag{18.32c}$$

$$R^\phi{}_{\theta\phi\theta} = -\frac{2m}{r^3} \tag{18.32d}$$

from which it is evident that the curvature diverges as $r \to 0$.

18.9 Geodesics in the Schwarzschild metric

The general equation for a geodesic is

$$\frac{d^2x^i}{ds^2} + \Gamma^i{}_{jk}\frac{dx^j}{ds}\frac{dx^k}{ds} = 0$$

where s is the geodesic's affine parameter. We may use τ or a linear function of τ as said parameter:

$$s = a\tau + b \tag{18.33}$$

From the expressions for $\Gamma^i{}_{jk}$ in the Schwarzschild metric, we have

$$\frac{d^2r}{ds^2} + \frac{A'}{2A}\left(\frac{dr}{ds}\right)^2 - \frac{r}{A}\left(\frac{d\theta}{ds}\right)^2 - \frac{r\sin^2\theta}{A}\left(\frac{d\phi}{ds}\right)^2 + \frac{B'}{2A}\left(\frac{dt}{ds}\right)^2 = 0 \tag{18.34}$$

$$\frac{d^2\theta}{ds^2} + \frac{2}{r}\frac{d\theta}{ds}\frac{dr}{ds} - \sin\theta\cos\theta\left(\frac{d\phi}{ds}\right)^2 = 0 \tag{18.35}$$

$$\frac{d^2\phi}{ds^2} + \frac{2}{r}\frac{d\phi}{ds}\frac{dr}{ds} + 2\cot\theta\frac{d\phi}{ds}\frac{d\theta}{ds} = 0 \tag{18.36}$$

$$\frac{d^2t}{ds^2} + \frac{B'}{B}\frac{dt}{ds}\frac{dr}{ds} = 0 \tag{18.37}$$

We have spherical symmetry, so without loss of generality we may consider the geodesic to lie in the hyperplane $\theta = \frac{\pi}{2}$ (recall that there is no torsion and thus the curve lies in a plane). With this choice, all the terms $\frac{d\theta}{ds}$ disappear and equation (18.35) is reduced to an identity.

From equation (18.37), dividing by $\frac{dt}{ds}$, we have

$$\frac{d}{ds}\left[\ln\frac{dt}{ds} + \ln B\right] = 0$$

Therefore,

$$\ln\left(B\frac{dt}{ds}\right) = \ln\frac{dt}{ds} + \ln B = k_1$$

(where k_1 is an integration constant), that is, $B\frac{dt}{ds} = e^{k_1}$, and choosing $k_1 = 0$ (which fixes the affine parameter s) we obtain

$$\frac{dt}{ds} = \frac{1}{B} \tag{18.38}$$

From equation (18.36), dividing by $\frac{d\phi}{ds}$, we have

$$\frac{d}{ds}\left[\ln\frac{d\phi}{ds} + \ln r^2\right] = 0$$

and, similarly,

$$r^2\frac{d\phi}{ds} = k_2 \tag{18.39}$$

(where k_2 is an integration constant).

Substituting equations (18.38) and (18.39) into equation (18.34) gives us

$$\frac{d^2r}{ds^2} + \frac{A'}{2A}\left(\frac{dr}{ds}\right)^2 - \frac{k_2^2}{Ar^3} + \frac{B'}{2AB^2} = 0$$

and, multiplying by $2A\frac{dr}{ds}$, we obtain

$$\frac{d}{ds}\left[\left(A\frac{dr}{ds}\right)^2 + \frac{k_2^2}{r^2} - \frac{1}{B}\right] = 0$$

that is,

$$A\left(\frac{dr}{ds}\right)^2 + \frac{k_2^2}{r^2} - \frac{1}{B} = k_3 \tag{18.40}$$

(where k_3 is an integration constant).

Using equations (18.39) and (18.40) in the expression for the Schwarzschild metric with $\theta = \frac{\pi}{2}$, we have

$$d\tau^2 = -k_3 \, ds^2 \qquad (18.41)$$

that is, $\frac{d\tau}{ds} = const.$, as we would expect for an affine parameter.

Note the following:

(i) $k_3 = 0 \Rightarrow d\tau^2 = 0 \Rightarrow$ the geodesic is null, that is, it is the trajectory a photon would follow.

(ii) $k_3 > 0 \Rightarrow d\tau^2 < 0 \Rightarrow$ the geodesic is time-like, that is, it is the trajectory a particle with nonzero mass would follow.

(iii) From equation (18.39), k_2 is the 'angular momentum per unit mass'.

(iv) Eliminating ds from equations (18.39) and (18.40), we obtain

$$\frac{A}{r^4}\left(\frac{dr}{d\phi}\right)^2 + \frac{1}{r^2} - \frac{1}{Bk_2^2} - \frac{k_3}{k_2^2} = 0 \qquad (18.42)$$

which gives us the orbit $r = r(\phi)$ of the geodesic.

18.10 Comments on Einstein's equations

In section 18.7 we obtained the field equations relating the gravitational field (given by the metric g_{ab} of the space–time M) with the distribution of matter on M (determined by an energy–momentum tensor T_{ab}; 'matter' here means 'anything that is not gravity', that is, all external fields, including electromagnetic ones).

These equations, called 'Einstein's equations', are of the form $R_{ab} - \frac{1}{2}Rg_{ab} = 8\pi T_{ab}$ and constitute a system of ten coupled nonlinear differential equations (since both sides of the equations are symmetric) for the components of the metric g_{ab} and their first and second derivatives, $g_{ab,c}$ and $g_{ab,cd}$.

However, only six of these equations are linearly independent, since

$$T^{ab}{}_{;b} = 0$$

and thus

$$\left(R^{ab} - \frac{1}{2}Rg^{ab}\right)_{;b} = 0$$

This is precisely the correct number of equations for (M, g_{ab}), since the four degrees of freedom we enjoy to carry out coordinate transformations allow us to fix four of the ten components of g_{ab}, leaving only six to determine.

On the other hand, any Lorentzian metric satisfies Einstein's equations: we simply *define* T_{ab} as the left-hand side of the equations divided by 8π. The matter described by such a tensor, however, will generally have irrational physical properties. By 'an exact solution to Einstein's equations', we wish to indicate a space–time (M, g_{ab}) with Lorentzian metric satisfying Einstein's equations for an energy–momentum tensor T_{ab} which describes a reasonable matter content.

What exactly does 'reasonable' mean?

Hawking and Ellis in *The Large Scale Structure of Space–Time* write the energy–momentum tensor in four canonical ways:

(i) If T_{ab} has a time-like eigenvector e_0, the tensor may be written as

$$T_{ab} = \begin{pmatrix} \mu & 0 & 0 & 0 \\ 0 & p_1 & 0 & 0 \\ 0 & 0 & p_2 & 0 \\ 0 & 0 & 0 & p_3 \end{pmatrix} \tag{18.43}$$

with respect to an orthonormal basis $\{e_0, e_1, e_2, e_3\}$ at a point p. In this case μ is the energy density for an observer whose worldline has e_0 as its tangent vector at p, and p_α are the principal pressures in the directions e_α ($\alpha \in \{1, 2, 3\}$). All known fields with nonzero rest mass have an energy–momentum tensor of this form; so do all fields with zero rest mass which do not have a double null eigenvector. The given form for T_{ab} indicates that its four eigenvalues are real. Therefore, its four eigenvectors are also real. Because they are mutually perpendicular, one of them must necessarily be time-like and the other three must necessarily be space-like (because no two real null vectors can be perpendicular to each other).

(ii) If T_{ab} has a double null eigenvector $e_0 + e_1$ (using the above notation) it may be written as

$$T_{ab} = \begin{pmatrix} \nu + \kappa & -\nu & 0 & 0 \\ -\nu & \nu - \kappa & 0 & 0 \\ 0 & 0 & p_2 & 0 \\ 0 & 0 & 0 & p_3 \end{pmatrix} \tag{18.44}$$

with respect to $\{e_0, e_1, e_2, e_3\}$. The only known field with such an energy–momentum tensor is the one represented by electromagnetic radiation travelling in the direction of $e_0 + e_1$. In this case, $p_2 = p_3 = \kappa = 0$. If T_{ab} has a null eigenvector n (i.e. $n^i n_i = 0$), then, assuming that n lies on the plane defined by e_0 and e_1, the equation $T_{ab}n^b = \kappa n_a$ gives

$$T_{00} + T_{01} = \kappa$$
$$T_{10} + T_{11} = -\kappa$$

Therefore, defining $T_{01} = -\nu$, we are left with $T_{00} = \nu + \kappa$, $T_{11} = \nu - \kappa$. Choosing the axes e_2 and e_3 along the two space-like eigenvectors, we obtain the given form of T_{ab}. Solving the equation $|T_{ab} - \lambda g_{ab}| = 0$, it is clear that two eigenvalues are equal: $\lambda_0 = \lambda_1 = -\kappa$, $\lambda_2 = p_2$, $\lambda_3 = p_3$. This is why n is called a 'double (null) eigenvector'.

(iii) If T_{ab} has a triple null eigenvector $e_0 + e_1$, it takes the form

$$T_{ab} = \begin{pmatrix} \nu & 0 & 1 & 0 \\ 0 & -\nu & 1 & 0 \\ 1 & 1 & -\nu & 0 \\ 0 & 0 & 0 & p \end{pmatrix} \tag{18.45}$$

In this case, T_{ab} has three degenerate eigenvalues; in fact, $\lambda_0 = \lambda_1 = \lambda_2 = \nu$, $\lambda_3 = p$. There are no known fields with a energy–momentum tensor like this.

(iv) Finally, if T_{ab} has two real eigenvalues and two complex ones then it looks like

$$T_{ab} = \begin{pmatrix} \nu & \kappa & 0 & 0 \\ \kappa & -\nu & 0 & 0 \\ 0 & 0 & p_2 & 0 \\ 0 & 0 & 0 & p_3 \end{pmatrix} \tag{18.46}$$

Suppose the equation $|T_{ab} - \lambda g_{ab}| = 0$ has two real solutions ($\lambda_2 = p_2$, $\lambda_3 = p_3$) and two complex solutions ($\lambda_{0,1} = \nu \pm i\kappa$). Let n be the eigenvector corresponding to $\lambda_0 = \nu + i\kappa$ and $\bar{n}$ be the one corresponding to $\lambda_1 = \nu - i\kappa$, and let us write

$$n_i = u_i + iv_i$$
$$\bar{n}_i = u_i - iv_i$$

As n_i and $\bar{n}_i$ are defined up to a multiplicative factor, we may ask that they be normalised: $n_i n^i = \bar{n}_i \bar{n}^i = 1$. Since $n_i \bar{n}^i = 0$ (because the two vectors are perpendicular to each other),

$$u_i u^i + v_i v^i = 0$$
$$u_i v^i = 0$$
$$u_i u^i - v_i v^i = 1$$

whereby $u_i u^i = -v_i v^i = \frac{1}{2}$, that is, one of these vectors (v) is time-like and the other (u) is space-like. Taking the coordinate axes along v, u, and the two eigenvectors with real eigenvalues, T_{ab} acquires the given form. Again, no known fields have such an energy–momentum tensor.

T_{ab} cannot have two conjugated pairs of eigenvalues, since only one of its eigenvectors can be time-like.

In general, the energy–momentum tensor for a real case will comprise contributions from several matter fields, so it will be extremely complicated (if not outright impossible) to describe. However, there are certain conditions on T_{ab} which are reasonably physical:

(i) *Weak energy condition.*

This condition asks that for all $p \in M$ and for any time-like vector $X \in T_p(M)$ we have

$$T_{ab}X^aX^b \geqslant 0 \tag{18.47}$$

By continuity, this condition is extended to null vectors. For an observer whose worldline at p has a unit tangent vector Y_p, the local energy density is $T_{ab}Y_p^aY_p^b$. The weak energy condition, then, tells us that the local energy density measured by any observer must be nonnegative, which is physically reasonable. In fact, the condition is satisfied for all experimentally detected fields.

(ii) *Dominant energy condition.*

This condition asks that for any time-like vector X we have

$$T^{ab}X_aX_b \geqslant 0$$

and that $T^{ab}X_a$ not be space-like. In other words, the dominant condition is the weak condition plus the requirement that the local energy flux measured by any observer be a time-like or null vector. This condition is satisfied by all known matter fields.

(iii) *Strong energy condition.*

This condition asks that for any time-like vector X we have

$$T_{ab}X^aX^b \geqslant \frac{1}{2}X^aX_aT$$

with $T = \mathrm{tr}(T_{ab})$, which is also physically reasonable. This would be violated only in the presence of a negative energy density or an extremely high negative pressure.

(iv) *Local causality condition.*

The equations governing matter fields must be such that a causality principle is satisfied locally, that is, such that, given a neighbourhood $U \subset M$ and two points $p, q \in U$, a signal can be sent from p to q only if there is a curve $\gamma: \mathbb{R} \longrightarrow U$ such that $\gamma(0) = p$, $\gamma(1) = q$ and $\dot{\gamma}(t)$ is timelike or null.

In general, we consider an energy–momentum tensor T_{ab} describing the matter content of the Universe to be 'physically reasonable' if it satisfies the local causality condition and at least one of the energy conditions.

One then seeks solutions to Einstein's equations which involve these kinds of energy–momentum tensors. Given the equations' complexity, only a few exact solutions, all of them for highly symmetric or very simple (in terms of the matter content) configurations (perfect fluids, electromagnetic radiation, etc), are known. However, all of these solutions give us qualitative ideas of the properties of the real universe.

Chapter 19

Gravitational radiation

19.1 Linear theory of gravitation

After decades of efforts, gravitational waves, general relativity's prediction that perturbations to the metric propagate through space–time in the form of waves, were finally discovered by the Laser Interferometer Gravitational-Wave Observatory (LIGO), an international collaboration which, at the time of writing, has two interferometric detectors which bounce laser beams between mirrors at opposite ends of perpendicularly set 4 km long vacuum tubes. When a gravitational wave passes through the detector it slightly alters the length of the tubes, stretching one while contracting the other in an oscillatory manner, thus producing a change of phase in the interference pattern of the beams.

The first signal observed simultaneously by both detectors, corresponds to the collision of two black holes, one with a mass of 36 and the other of 29 solar masses, giving rise to a final 62 solar mass black hole; the remaining 3 solar masses is what was radiated as gravitational waves. (Other detectors, like the Advanced Virgo Detector, are now working and several other events have been detected, and there are still many more to come.)

Prior to this first event there was indirect evidence of the existence of gravitational radiation from radio flashes emitted by pairs of neutron stars orbiting each other. As the waves carry energy away from the system, the timing of the flashes shifts, and these shifts were in accordance with the predictions of general relativity.

Like in electrodynamics, gravitational interactions travel at the speed of light across the manifold M giving rise to gravitational radiation. Since this radiation is very weak, we will study it as a first-order perturbation to the metric. We will also calculate the energy radiated by a gravitational wave.

Consider

$$g_{ab} = \eta_{ab} + h_{ab} \qquad (19.1)$$

doi:10.1088/2053-2563/aadf65ch19

where $h_{ab} \ll 1$, and let us work at first order in h_{ab}. The energy–momentum tensor satisfies $T^{ab}_{,b} = 0$, as in special relativity, and, since there is no covariant derivative in that last expression, in the linear theory the gravitational field does not exert any influence on the movement of the matter producing it.

Calculating the Riemann tensor for the metric of equation (19.1), we obtain

$$R^a_{mbn} = \frac{1}{2}\eta^{ai}(h_{in,\, mb} + h_{mb,\, in} - h_{mn,\, bi} - h_{bi,\, mn}) \qquad (19.2)$$

to first order in h_{ab}, whereby

$$8\pi T_{mn} = R_{mn} - \frac{1}{2}R\,\eta_{mn}$$

$$= -\frac{1}{2}\left(h_{mn}{}^{,a}{}_{,a} + h^a{}_{a,\, mn} - h^a{}_{n,\, ma} - h^a{}_{m,\, na} - \eta_{mn}h^i{}_i{}^{,a}{}_{,a}\right) \qquad (19.3)$$

$$+ \eta_{mn}h^{ab}{}_{,ab}\Big)$$

Defining

$$\bar{h}_{mn} = h_{mn} - \frac{1}{2}h\,\eta_{mn} \qquad (19.4)$$

with $h = h^a{}_a$, equation (19.3) is reduced to

$$\bar{h}_{mn}{}^{,a}{}_{,a} + \eta_{mn}\bar{h}^{ab}{}_{,ab} - \bar{h}^a{}_{n,\, ma} - \bar{h}^a{}_{m,\, na} = -16\pi T_{mn} \qquad (19.5)$$

Performing a coordinate transformation

$$\hat{x}^a = x^a + \xi^a(x)$$

whereby

$$\widehat{\bar{h}}^{mn} = \bar{h}^{mn} - \xi^{n,\, m} - \xi^{m,\, n} + \eta^{mn}\xi^a{}_{,a}$$

we may set (by choosing ξ^a such that $\Box\xi^a = \eta^{ij}\xi^a{}_{,ij} = \bar{h}^{ba}{}_{,a}$)

$$\bar{h}^{mn}{}_{,n} = \left(\sqrt{-g}\,g^{mn}\right)_{,n} = 0 \qquad (19.6)$$

Equation (19.5) for the perturbation of the metric is then reduced to

$$\Box\,\bar{h}_{mn} = -16\pi T_{mn} \qquad (19.7)$$

which is a wave equation, that is, the perturbations of η_{mn} propagate across M as a wave, and they do so at the speed of light. Here, $\Box = \nabla_m\nabla^m = g^{mn}\nabla_m\nabla_n$ is the d'Alembert operator, with g^{mn} the inverse metric tensor and ∇ the metric connection in M.

At higher orders in h_{mn}, $\bar{h}_{mn}$ also satisfies a wave equation, but one with terms proportional to powers of h_{mn} on the right-hand side of equation (19.7). In other words, even in the absence of an energy–momentum tensor, $\bar{h}_{,m}$ satisfies a 'forced wave equation' where the forcing term is h_{mn} itself (this is a consequence of the nonlinearity of the theory).

If we consider perturbations to a non-flat metric $g_{ab}^{(0)}$, we find that

$$\Gamma^i{}_{jk}{}^{(1)} = \frac{1}{2}\left(h^i{}_{k;l} + h^i{}_{l;k} - h_{kl}{}^{;i}\right)$$

whereby

$$R^i{}_{klm}{}^{(1)} = \frac{1}{2}\left(h^i{}_{k;ml} + h^i{}_{m;kl} - h_{km}{}^{;i}{}_{;l} - h^i{}_{k;lm} - h^i{}_{l;km} + h_{kl}{}^{;i}{}_{;m}\right)$$

Therefore, since $g_{ab}^{(0)}$ satisfies $R_{ab}^{(0)} = 0$, in vacuum ($R_{ik} = 0$) $g_{ab}^{(1)}$ will satisfy $R_{ab}^{(1)} = 0$, that is, contracting the above equation we obtain

$$h_{ik}{}^{;l}{}_{;l} - h^l{}_{i;kl} - h^l{}_{k;il} + h_{;ik} = 0$$

which is a wave equation like equation (19.3) but cannot, in general, be reduced to the simplified form seen in equation (19.7).

19.2 Comparison with electromagnetism

	Electromagnetism	General relativity: linear theory
Fundamental variable	A_m	$\bar{h}_{mn}$
Field equations	$A_m{}^{,a}{}_{,a} - A^a{}_{,ma} = -J_m$	equation (19.5)
Equations invariant under	$\hat{A}_m = A_m + \xi_{,m}$	$\hat{\bar{h}}^{mn} = \bar{h}^{mn} - \xi^{n,m} - \xi^{m,n} + \eta^{mn}\xi^a{}_{,a}$
	(gauge transformation)	(coordinate transformation)
Lorentz condition	$A^m{}_{,m} = 0$	$\bar{h}^{mn}{}_{,n} = 0$
Simplified field equations	$\Box A_m = -J_m$	$\Box \bar{h}_{mn} = -16\pi T_{mn}$
Remaining gauge Freedom	$\Box \xi = 0$	$\Box \xi^m = 0$

19.3 Solutions to the wave equation

The great analogy between electrodynamics and the linearised version of Einstein's theory allows us (see equation (19.7)) to also write solutions to the field equations in terms of retarded potentials:

$$\bar{h}_{mn}(r,\ t) = \int \frac{T_{mn}(\bar{r},\ t - |r - \bar{r}|)}{|r - \bar{r}|} d^3\bar{x} \tag{19.8}$$

where the integral is taken over the sources.

A solution to the homogeneous equation $\Box \bar{h}_{mn} = 0$ can always be added to the particular solution of equation (19.8). In this way, we can switch between retarded and advanced potentials, for instance.

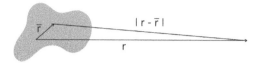

Figure 19.1. The field seen at large distances from a compact source.

If the mean speeds of the sources are small and have a compact support, then at large distances from the sources we can write the field as

$$\bar{h}_{mn} = \frac{1}{r} \int T_{mn}(\bar{r}, t - r)d^3\bar{x} \tag{19.9}$$

where r is the distance from the observation point to any point of the sources. This is depicted in figure 19.1.

Now, from $T^{ab}_{,b} = 0$ we have

$$\frac{\partial}{\partial x^\gamma}T^{\alpha\gamma} + \frac{\partial}{\partial x^0}T^{\alpha 0} = 0 \tag{19.10}$$

$$\frac{\partial}{\partial x^\gamma}T^{0\gamma} + \frac{\partial}{\partial x^0}T^{00} = 0 \tag{19.11}$$

Multiplying equation (19.10) by x^β and integrating over all of space,

$$\frac{\partial}{\partial x^0} \int T^{\alpha 0}x^\beta \, dV \quad = \quad -\int \frac{\partial}{\partial x^\gamma}T^{\alpha\gamma}x^\beta \, dV$$

$$= \quad -\int \frac{\partial}{\partial x^\gamma}(T^{\alpha\gamma}x^\beta)dV + \int T^{\alpha\beta} \, dV$$

$$\underset{\substack{\text{Gauss's theorem} \\ (T_{\alpha\beta}=0 \text{ at } \infty)}}{=} \int T^{\alpha\beta} \, dV$$

Therefore, since $T^{\alpha\beta}$ is symmetric,

$$\int T^{\alpha\beta} \, dV = \frac{1}{2}\frac{\partial}{\partial x^0} \int (T^{\alpha 0}x^\beta + T^{\beta 0}x^\alpha)dV \tag{19.12}$$

Multiplying equation (19.11) by $x^\alpha x^\beta$ and integrating, we obtain in the same way

$$\frac{\partial}{\partial x^0} \int T^{00}x^\alpha x^\beta \, dV = \int (T^{\alpha 0}x^\beta + T^{\beta 0}x^\alpha)dV \tag{19.13}$$

Comparing equations (19.12) and (19.13),

$$\int T^{\alpha\beta} \, dV = \frac{1}{2}\frac{\partial^2}{\partial(x^0)^2} \int T^{00}x^\alpha x^\beta \, dV \tag{19.14}$$

that is, all the integrals where $T^{\alpha\beta}$ is part of the integrand can be replaced by integrals where only $T^{00} = \rho$ appears. In particular (see equation (19.9)),

$$\bar{h}_{\alpha\beta}(r, t) = \frac{1}{2r} \frac{\partial^2}{\partial t^2} \int \rho x^\alpha x^\beta \, dV \tag{19.15}$$

For a plane wave moving in the direction of x^1, the only nonzero components of $\bar{h}_{ab}$ are $\bar{h}_{22}$, $\bar{h}_{23}$ and $\bar{h}_{33}$. Defining

$$D_{\alpha\beta} \overset{\text{def}}{=} \int \rho \big(3x_\alpha x_\beta - r^2 \delta_{\alpha\beta}\big) dV \tag{19.16}$$

which we call the 'quadrupolar mass moment', we obtain

$$\bar{h}_{22} = \frac{1}{6r} \ddot{D}_{22}$$

$$\bar{h}_{23} = \frac{1}{6r} \ddot{D}_{23}$$

$$\bar{h}_{33} = \frac{1}{6r} \ddot{D}_{33}$$

Note that gravitational waves are transverse waves: their polarisation tensor is a symmetric tensor in the $x^2 x^3$ plane. In general, one would expect for three polarisations to be present, $x^2 x^2$, $x^2 x^3$, and $x^3 x^3$, but the gauge freedom in the choice of reference frame given by ξ reduces these to two.

In general, in a space–time manifold of dimension d we will have

$$\# \text{ polarisations} = \frac{(d-1)(d-2)}{2} - 1$$

and this means that for $d = 2$ ($= 1 + 1$) and for $d = 3$ ($= 2 + 1$) there are no gravitational waves. Thus, solutions to Einstein's equations which are spherically symmetric or axis-symmetric do not radiate.

19.4 Energy of the gravitational field

In the absence of a gravitational field, the conservation of energy and momentum of matter fields is written as

$$T^{ab}_{,b} = 0 \tag{19.17}$$

However, in the presence of a gravitational field Bianchi's identities give us

$$T^{ab}_{;b} = 0 \tag{19.18}$$

which is not a conservation law. In other words, the energy and momentum of the matter fields are *not* conserved in the presence of gravitational fields: there is an exchange of these quantities between the matter fields and the gravitational field.

It is therefore natural to think that the conservation of the total energy and momentum (those of the matter fields and those of the gravitational fields) can be expressed as

$$(T^{ab} + t^{ab})_{,b} = 0$$

where t^{ab} measures the energy and momentum of the gravitational field itself; or more generally as

$$[(-g)^n(T^{ab} + t^{ab})]_{,b} = 0 \tag{19.19}$$

for some $n \in \mathbb{Z}$ (or such that $2n \in \mathbb{Z}$). It turns out that there are multiple expressions for t^{ab} which satisfy equation (19.19). In addition, an essential property of t^{ab} is that it is not a tensor, but rather a 'pseudo-tensor'. Among the various 'energy–momentum pseudo-tensors for the gravitational field', the Landau–Lifshitz pseudo-tensor t_{LL} has the property of being symmetric (and, therefore, besides defining energy and linear momentum, it allows us to define angular momentum) and also satisfies

$$\frac{\partial}{\partial x^k}[(-g)(T^{ik} + t_{LL}^{ik})] = 0 \tag{19.20}$$

It is given by

$$
\begin{aligned}
(-g)t_{LL}^{ik} = \frac{1}{16\pi}\Bigg[& \mathfrak{A}^{ik}{}_{,l}\mathfrak{A}^{lm}{}_{,m} - \mathfrak{A}^{il}{}_{,l}\mathfrak{A}^{km}{}_{,m} + \frac{1}{2}g^{ik}g_{lm}\mathfrak{A}^{ln}{}_{,p}\mathfrak{A}^{pm}{}_{,n} \\
& - \left(g^{il}g_{mn}\mathfrak{A}^{kn}{}_{,p}\mathfrak{A}^{mp}{}_{,l} + g^{kl}g_{mn}\mathfrak{A}^{in}{}_{,p}\mathfrak{A}^{mp}{}_{,l}\right) + g_{lm}g^{np}\mathfrak{A}^{il}{}_{,n}\mathfrak{A}^{km}{}_{,p} \\
& + \frac{1}{8}(2g^{il}g^{km} - g^{ik}g^{lm})\left(2g_{np}g_{qr} - g_{pq}g_{nr}\right)\mathfrak{A}^{nr}{}_{,l}\mathfrak{A}^{pq}{}_{,m} \Bigg]
\end{aligned}
\tag{19.21}
$$

where

$$\mathfrak{A}^{ik} \overset{\text{def}}{=} \sqrt{-g}\, g^{ik} \tag{19.22}$$

Several other energy–momentum pseudo-tensors exist, some with better properties than others (notably the 'Noether gravitational operator' introduced by Schutz and Sorkin).

19.5 Energy radiated by a gravitational wave

We wish to calculate the energy–momentum pseudo-tensor for a plane gravitational wave. Note that t^{ik} is a second-order function of h_{ab} (all of its terms in equation (19.21) have a product of two derivatives of g_{ik}); therefore, to calculate it we must ignore all higher-order terms. Because all of the terms in equation (19.21) are exactly of second order in h_{ab}, it does not matter that we have previously calculated everything to first order in h_{ab} only: any term of order higher than 1 in g_{ab} would result in terms of order higher than 2 in h_{ab} in the expression for t^{ab}.

$\det(g_{ab})$ differs from $\det(\eta_{ab}) = -1$ by terms of second order in h_{ab}. Therefore,

$$\mathfrak{Y}^{ik}{}_{,l} \approx g^{ik}{}_{,l} \approx -h^{ik}{}_{,l} \tag{19.23}$$

For a plane wave moving in the direction of x^1, we thus have, from equation (19.21) and the expressions for the quadrupolar mass moment,

$$t^{10} = \frac{1}{4\pi r^2}\left[\left(\frac{\ddot{D}_{22} - \ddot{D}_{33}}{2}\right)^2 + \ddot{D}_{23}^2\right] \tag{19.24}$$

This quantity measures the energy flux in the direction of x^1.

The energy per unit time which passes through a solid-angle element $d\Omega$ in the direction of x^1 is obtained by multiplying the above expression by $r^2\,d\Omega$:

$$\frac{dE}{dt}(x^1,\,d\Omega) = \frac{1}{4\pi}\left[\left(\frac{\ddot{D}_{22} - \ddot{D}_{33}}{2}\right)^2 + \ddot{D}_{23}^2\right]d\Omega \tag{19.25}$$

Here, $\left(\frac{\ddot{D}_{22} - \ddot{D}_{33}}{2}\right)^2$ is the contribution of the $(+)$ polarisation and $\ddot{D}_{23}^2$ is the contribution of the $(\times)$ polarisation. Note that, in an arbitrary direction $\vec{n}$, the contribution of any given polarisation is $\frac{1}{8\pi}(\ddot{D}_{\alpha\beta}e^{\alpha\beta})^2\,d\Omega$, where $e_{\alpha\beta}$ is the 'polarisation tensor' and is given by $e_{\alpha\beta} = e_{\beta\alpha}$, $e_{\alpha\alpha} = 0$, $e_{\alpha\beta}n_\beta = 0$, $e_{\alpha\beta}e_{\alpha\beta} = 1$. The angular distribution given by all polarisations is obtained by averaging over the polarisations and multiplying by the number of independent polarisations (in this case, two).

To obtain the total energy radiated in all directions (that is, the energy per unit time lost by the system), we average $\frac{dE}{dt\,d\Omega}$ over all directions and multiply by 4π.

$$\frac{dE}{dt} = \frac{1}{5}\left\langle \ddot{D}_{\alpha\beta}\ddot{D}^{\alpha\beta}\right\rangle \tag{19.26}$$

where $\langle\rangle$ here denotes an average over several characteristic periods of the source.

To average over polarisations, we use the expression

$$\langle e_{\alpha\beta}\,e_{\gamma\delta}\rangle = \frac{1}{4}\Big[n_\alpha n_\beta n_\gamma n_\delta + n_\alpha n_\beta \delta_{\gamma\delta} + n_\gamma n_\delta \delta_{\alpha\beta}$$
$$- \left(n_\alpha n_\gamma \delta_{\beta\delta} + n_\beta n_\gamma \delta_{\alpha\delta} + n_\alpha n_\delta \delta_{\beta\gamma} + n_\beta n_\delta \delta_{\alpha\gamma}\right)$$
$$- \delta_{\alpha\beta}\delta_{\gamma\delta} + \delta_{\alpha\gamma}\delta_{\beta\delta} + \delta_{\beta\gamma}\delta_{\alpha\delta}\Big]$$

which is obtained by asking that the right-hand side be a tensor formed by δ_{ik} and $\vec{n}$, that it be symmetric under $\alpha \leftrightarrow \beta$ and $\gamma \leftrightarrow \delta$, whose contraction of α with γ and of β with δ equals 1, and whose contraction with $\vec{n}$ vanishes.

To average the product of two and four $\vec{n}$'s over all the directions, we proceed as follows:

(i) $\langle n_\alpha n_\beta\rangle = \frac{1}{4\pi}\int n_\alpha n_\beta\,d\Omega$. Since $n_\alpha n_\beta$ is a symmetric unit tensor, we can write it in terms of $\delta_{\alpha\beta}$. Noting that it has unit trace, we then have

$$\langle n_\alpha n_\beta \rangle = \frac{1}{3}\delta_{\alpha\beta}$$

(ii) $\langle n_\alpha n_\beta n_\gamma n_\delta \rangle = \frac{1}{4\pi}\int n_\alpha n_\beta n_\gamma n_\delta \, d\Omega$. We must build a rank-$\begin{pmatrix} 0 \\ 4 \end{pmatrix}$ tensor from $\delta_{\alpha\beta}$, symmetric in all of its indices, and whose contraction on any two pairs of indices equals 1. The result is

$$\langle n_\alpha n_\beta n_\gamma n_\delta \rangle = \frac{2}{3}\big(\delta_{\alpha\beta}\delta_{\gamma\delta} + \delta_{\alpha\gamma}\delta_{\beta\delta} + \delta_{\alpha\delta}\delta_{\beta\gamma}\big)$$

19.6 Example

In order to have an idea of how weak gravitational radiation is, let us calculate the power radiated as gravitational waves by a basketball being bounced by a player as he crosses the court.

The radiated power is given by equation (19.26). To obtain an estimate of its order of magnitude, we take the quadrupolar mass moment as $D_{\alpha\beta} = \int \rho x_\alpha x_\beta \, dV$.

The ball moves approximately as a harmonic oscillator: $x_\alpha = A_\alpha \sin(\omega t)$, where A_α is the amplitude of the motion and ω is its angular frequency. Hence, if m is the ball's mass,

$$D_{\alpha\beta} = A_\alpha A_\beta \sin^2(\omega t) \int \rho \, d^3x = m A_\alpha A_\beta \sin^2(\omega t)$$

and

$$\dddot{D}_{\alpha\beta} = -4m A_\alpha A_\beta \omega^3 \sin(2\omega t)$$

Therefore, averaging over one period,

$$\langle \dddot{D}_{\alpha\beta}\dddot{D}^{\alpha\beta} \rangle = 16m^2|A|^4\omega^6\frac{\omega}{\pi}\int_0^{\frac{\pi}{\omega}} \sin^2(2\omega t)dt$$

$$= 8m^2|A|^4\omega^6$$

The radiated power throughout one period is then

$$P = \frac{1}{5}8m^2|A|^4\omega^6 = \frac{8}{5}\frac{G}{c^5}m^2|A|^4\omega^6$$

where we have converted the expression into standard units in the last of the above equalities: $G \approx 6.7 \times 10^{-11}\text{N m}^2\text{kg}^{-2}$ is the universal gravitational constant, and $c \approx 3 \times 10^8 \text{ ms}^{-1}$ is the speed of light in a vacuum. Setting $m \approx 0.5$ kg, $|A| \approx 1$ m, and $\omega = \frac{2\pi}{\tau} \approx \frac{2\pi}{0.5 \text{ s}} = 4\pi\text{s}^{-1}$, we have

$$P \approx \frac{8}{5}\frac{6.7 \times 10^{-11}}{3^5 \times 10^{40}}\frac{1}{4}4^6\pi^6\frac{\text{N m}^2 \text{ s}^5 \text{ kg}^2 \text{ m}^4}{\text{kg}^2 \text{ m}^5 \text{ s}^6} \approx 4 \times 10^{-47}\text{W}$$

IOP Publishing

Differential Topology and Geometry with Applications to Physics

Eduardo Nahmad-Achar

Further reading

The following works will help the reader to delve further in some topics, or see them treated in a different way as presented in this volume. Naturally, the number of references that we can cite is huge, so we have been forced to select a few of them.

Many of the works mentioned cover more than one discipline. To help the reader, we have classified them under one discipline in accordance to the general orientation of the work.

Differential topology

[1] Wallace A 2010 *Differential Topology: First Steps* (New York: Dover)
[2] Milnor J W 1997 *Topology from the Differentiable Viewpoint* (Princeton, NJ: Princeton University Press)
[3] Singer I M and Thorpe J A 1976 *Lecture Notes on Elementary Topology and Geometry* (Berlin: Springer)
[4] Bröcker T B and Jänich K 1982 *Introduction to Differential Topology* (Cambridge: Cambridge University Press)

Differential geometry

[1] O'Neill B 2014 *Elementary Differential Geometry* (New York: Academic)
[2] Spivak M 1998 *Calculus on Manifolds* (Boulder, CO: Westview)
[3] Spivak M 1999 *A Comprehensive Introduction to Differential Geometry* vols 1 and 2 (Houston, TX: Publish or Perish)
[4] Boothby W M 2002 *An Introduction to Differentiable Manifolds and Riemannian Geometry* (New York: Academic)
[5] Choquet-Bruhat Y, DeWitt-Morette C and Dillard-Bleick M 2004 *Analysis, Manifolds and Physics* (Amsterdam: North Holland)
[6] Isham C J 1999 *Modern Differential Geometry for Physicists* (Singapore: World Scientific)
[7] Schutz B F 1980 *Geometrical Methods of Mathematical Physics* (Cambridge: Cambridge University Press)

General relativity

[1] Schutz B F 2009 *A First Course in General Relativity* (Cambridge: Cambridge University Press)
[2] Misner C W, Thorne K S and Wheeler J A 1973 *Gravitation* (San Francisco, CA: Freeman)
[3] Stephani H 1990 *General Relativity: An Introduction to the Theory of the Gravitational Field* (Cambridge: Cambridge University Press)
[4] Wald R M 2010 *General Relativity* (Chicago, IL: University of Chicago Press)
[5] Hawking S W and Ellis G F R 1975 *The Large Scale Structure of Space-Time* (Cambridge: Cambridge University Press)
[6] Rindler W 2006 *Relativity: Special, General, and Cosmological* (Oxford: Oxford University Press)
[7] Landau L D and Lifshitz E M 2000 *The Classical Theory of Fields: Volume 2 (Course of Theoretical Physics Series)* (Oxford: Butterworth Heinemann)

Gravitational radiation

[1] Castelvecchi D and Witze A 2016 Einstein's gravitational waves found at last *Nature News* www.nature.com/news/einstein-s-gravitational-waves-found-at-last-1.19361
[2] Abbott B P *et al* 2016 Observation of gravitational waves from a binary black hole merger *Phys. Rev. Lett.* **116** 061102

www.ingramcontent.com/pod-product-compliance
Ingram Content Group UK Ltd.
Pitfield, Milton Keynes, MK11 3LW, UK
UKHW050444181224
452686UK00003B/14